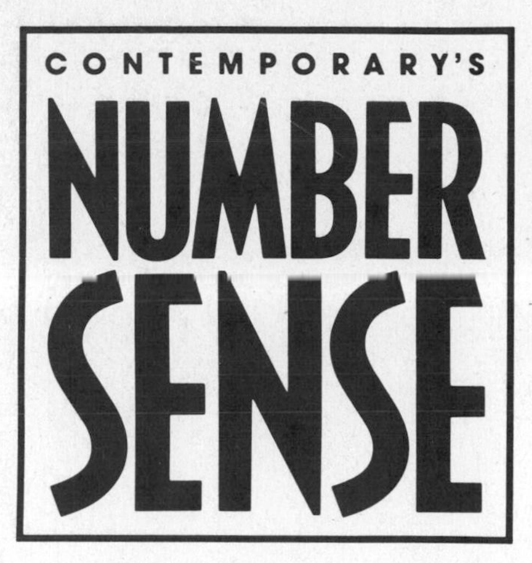

Discovering Basic Math Concepts

Ratio & Proportion

Allan D. Suter

Project Editors
Kathy Osmus
Caren Van Slyke

CB
CONTEMPORARY BOOKS
a division of NTC/Contemporary Publishing Group
Lincolnwood, Illinois USA

ISBN: 0-8092-4222-2

Published by Contemporary Books,
a division of NTC/Contemporary Publishing Group, Inc.,
4255 West Touhy Avenue,
Lincolnwood (Chicago), Illinois 60712-1975 U.S.A.

Manufactured in the United States of America.

0 1 2 3 4 5 6 7 C(K) 23 22 21 20 19 18 17

Editorial Director
Caren Van Slyke

Editorial
Seija Suter
Sarah Conroy
Ellen Frechette
Steve Miller
Robin O'Connor
Cliff Wirt
Jane Samuelson

Editorial Production Manager
Norma Fioretti

Production Editor
Craig Bolt

Cover Design
Lois Koehler

Illustrator
Ophelia M. Chambliss-Jones

Art & Production
Andrea Haracz

Typography
Impressions, Inc.
Madison, Wisconsin

Cover photo © C. C. Cain Photography.

Dedicated to our friend, Pat Reid

Contents

5 Proportion Applications

6 Proportion Problem Solving

What Is Ratio?

In your work with math, you may need to compare one number with another number.

One way to compare numbers is to use a **ratio.**

Compare the number of circles to the number of triangles.

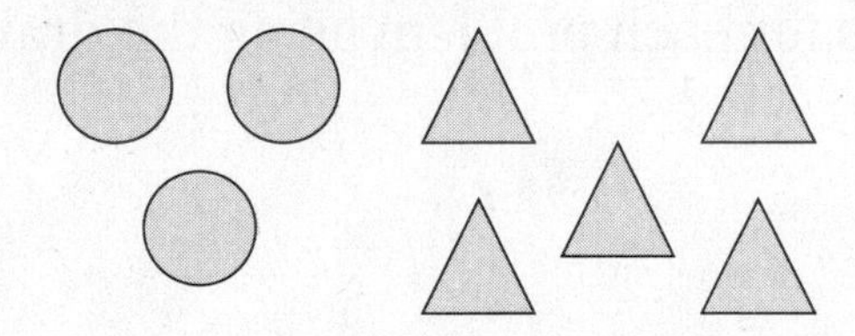

Ratios may be shown in three ways. All three ratios are read as "3 to 5." The numbers in a ratio are called **terms.**

3 to 5 $\qquad$ 3 : 5 $\qquad$ $\frac{3}{5}$

---

1. Compare the number of triangles to the number of circles.

   a) 5 to 3 $\qquad$ b) ____ : ____ $\qquad$ c) $\frac{\square}{\square}$

2. Compare the number of circles to the total number of figures.

   a) ____ to ____ $\qquad$ b) ____ : ____ $\qquad$ c) 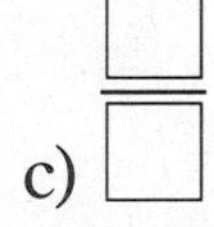

3. Compare the number of triangles to the total number of figures.

   a) ____ to ____ $\qquad$ b) ____ : ____ $\qquad$ c) $\frac{\square}{\square}$

Write the Ratios

Write the three forms of ratios for each problem using the drawings above.

1. Number of circles to number of squares — a) 2 to 3 b) ___ : ___ c) $\frac{\square}{\square}$

2. Number of squares to number of circles — a) ___ to ___ b) ___ : ___ c) $\frac{\square}{\square}$

3. Number of circles to total number of figures — a) ___ to ___ b) ___ : ___ c) $\frac{\square}{\square}$

4. Number of squares to total number of figures — a) ___ to ___ b) ___ : ___ c) $\frac{\square}{\square}$

5. Total number of figures to circles — a) ___ to ___ b) ___ : ___ c) 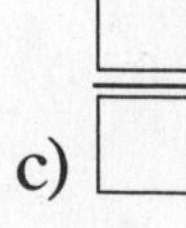

Ratios as Fractions

It is easier to solve ratio problems when you write the ratios as fractions.

Write the ratio as a fraction.

1. to

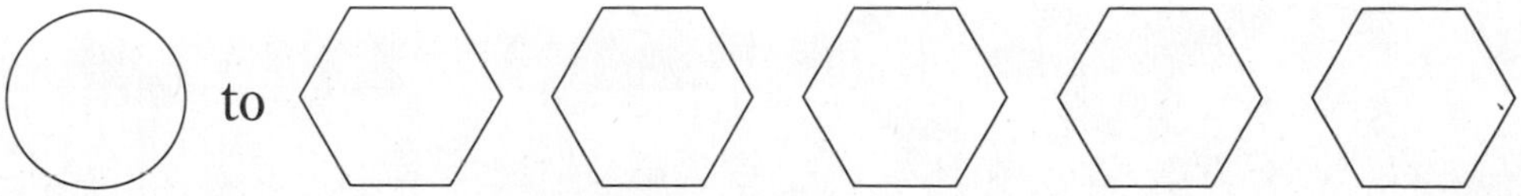

Compare circle to hexagons.

2.

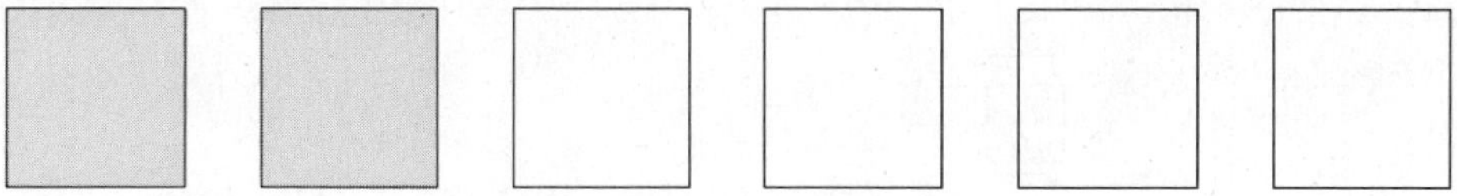

Show the ratio of shaded squares to total squares.

3. to

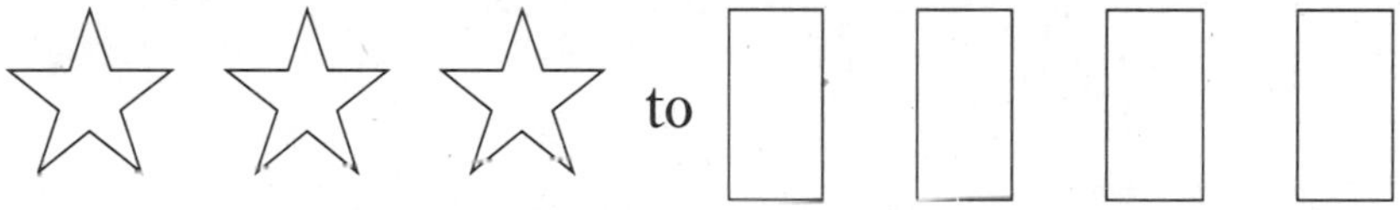

Compare stars to rectangles.

4.

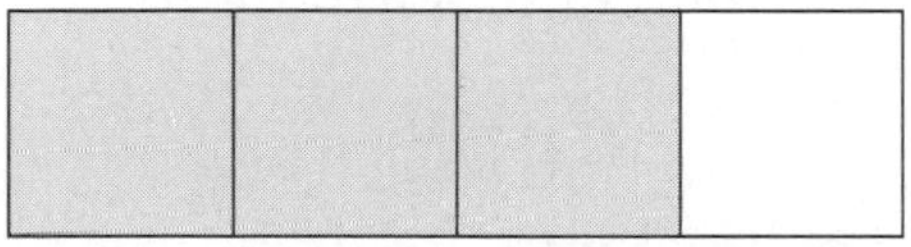

Compare the number of unshaded squares to shaded squares.

5. to

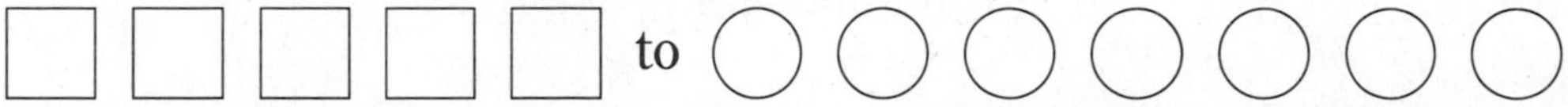

Compare squares to circles.

6.

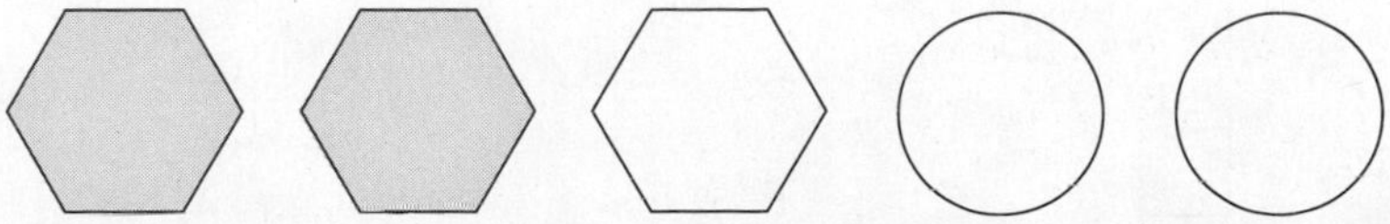

Compare shaded shapes to all shapes.

Compare the Shapes

Use the drawings above to complete the problems.

1. ___ : ___ = $\frac{\square}{\square}$ Compare the number of circles to the number of triangles.

2. ___ : ___ = $\frac{\square}{\square}$ Compare the number of triangles to the number of circles.

3. ___ : ___ = $\frac{\square}{\square}$ Compare the number of circles to the total number of figures.

4. ___ : ___ = $\frac{\square}{\square}$ Compare the number of triangles to the total number of figures.

5. ___ : ___ = $\frac{\square}{\square}$ Compare the total number of figures to the circles.

6. ___ : ___ = $\frac{\square}{\square}$ Compare the total number of figures to the triangles.

Write a ratio for each of the following problems.

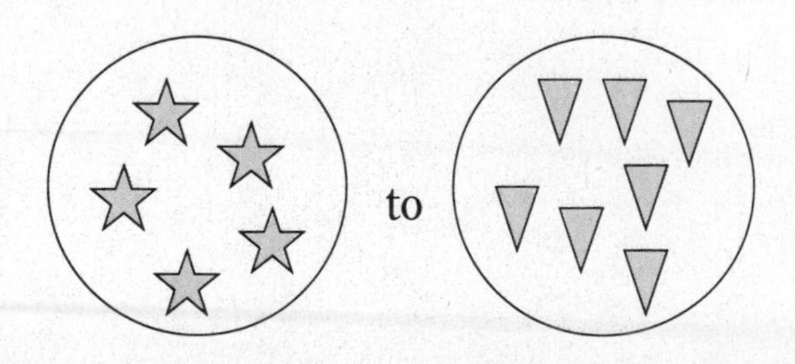

7. ___ : ___ = $\frac{\square}{\square}$

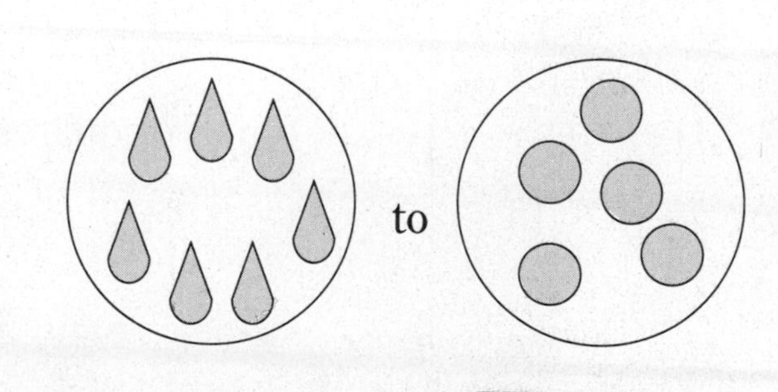

8. ___ : ___ = $\frac{\square}{\square}$

Draw the Ratios

1. Draw six Xs inside the rectangle. Circle one out of every three Xs.

Write circled Xs to total Xs as a fraction.

2. Draw eight squares inside the rectangle. Shade three out of every four squares.

Write shaded squares to total squares.

In comparing ratios, it is very important to keep the numbers in the same order as they are written.

Draw a picture for each ratio.

3. The ratio of stars to triangles is 3 to 5.

stars to triangles

stars
triangles

4. The ratio of triangles to stars is 5 to 3.

triangles to stars

triangles
stars

Write the Ratios as Fractions

Write a fraction to show how the first number compares with the second number.

1. $\frac{17}{30}$ There are 17 bicycles and 30 students.

2. $\frac{\square}{\square}$ There are 2 bicycles for every 3 students.

3. $\frac{\square}{\square}$ Sam paid $10 for 2 tickets.

4. $\frac{\square}{\square}$ Rose drove 384 miles in 7 hours.

5. $\frac{\square}{\square}$ The wall had 5 feet of width for every 12 feet of height.

6. $\frac{\square}{\square}$ The store sells 100 tablets of aspirin for $.89.

7. $\frac{\square}{\square}$ Craig bought 3 ties for $10.

8. $\frac{\square}{\square}$ The box was 6 inches wide and 10 inches long.

Simplify the Ratios

When a ratio is written as a fraction, it is usually simplified.

$$4 \text{ to } 32 = \frac{4}{32} \div \frac{4}{4} = \frac{1}{8}$$

Write each ratio as a fraction and simplify.

1. 5 to 20 $= \frac{5}{20}$ (simplify) $= \frac{\square}{\square}$
2. $12 : 15 = \frac{\square}{\square} = \frac{\square}{\square}$
3. 15 to 30 $= \frac{\square}{\square} = \frac{\square}{\square}$
4. 4 to 12 $= \frac{\square}{\square} = \frac{\square}{\square}$
5. $8 : 64 = \frac{\square}{\square} = \frac{\square}{\square}$
6. 6 to 8 $= \frac{\square}{\square} = \frac{\square}{\square}$
7. 8 to 24 $= \frac{\square}{\square} = \frac{\square}{\square}$
8. 16 to 18 $= \frac{\square}{\square} = \frac{\square}{\square}$
9. 9 to 36 $= \frac{9}{36}$ (simplify) $= \frac{\square}{\square}$
10. $21 : 24 = \frac{\square}{\square} = \frac{\square}{\square}$
11. $7 : 49 = \frac{\square}{\square} = \frac{\square}{\square}$
12. 6 to 18 $= \frac{\square}{\square} = \frac{\square}{\square}$
13. 28 to 35 $= \frac{\square}{\square} = \frac{\square}{\square}$
14. 10 to 25 $= \frac{\square}{\square} = \frac{\square}{\square}$
15. $11 : 22 = \frac{\square}{\square} = \frac{\square}{\square}$
16. 5 to 20 $= \frac{\square}{\square} = \frac{\square}{\square}$

Denominator of 1

When a ratio is written as a fraction, it must have a denominator.

$$6 : 2 = \frac{6}{2} \div \frac{2}{2} = \frac{3}{1}$$ ← The denominator is 1.

Simplifying a ratio does not change its value.

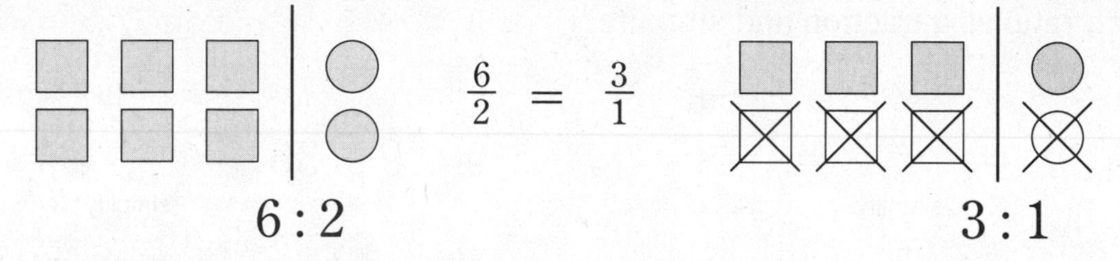

Write each ratio as a fraction and simplify.

1. 8 : 2 $= \frac{8}{2} = \frac{4}{1}$
2. 9 : 3 $= \frac{\square}{\square} = \frac{\square}{\square}$
3. 15 : 5 $= \frac{\square}{\square} = \frac{\square}{\square}$
4. 36 : 6 $= \frac{\square}{\square} = \frac{\square}{\square}$
5. 50 to 10 $= \frac{\square}{\square} = \frac{\square}{\square}$
6. 24 : 4 $= \frac{\square}{\square} = \frac{\square}{\square}$
7. 60 to 10 $= \frac{\square}{\square} = \frac{\square}{\square}$
8. 56 to 8 $= \frac{56}{8} = \frac{\square}{\square}$
9. 12 : 4 $= \frac{\square}{\square} = \frac{\square}{\square}$
10. 63 to 9 $= \frac{\square}{\square} = \frac{\square}{\square}$
11. 70 : 7 $= \frac{\square}{\square} = \frac{\square}{\square}$
12. 64 : 8 $= \frac{\square}{\square} = \frac{\square}{\square}$
13. 49 to 7 $= \frac{\square}{\square} = \frac{\square}{\square}$
14. 100 : 5 $= \frac{\square}{\square} = \frac{\square}{\square}$

Equal Ratios Are Equal Fractions

For every one X there are two circles. The ratio is 1 to 2.

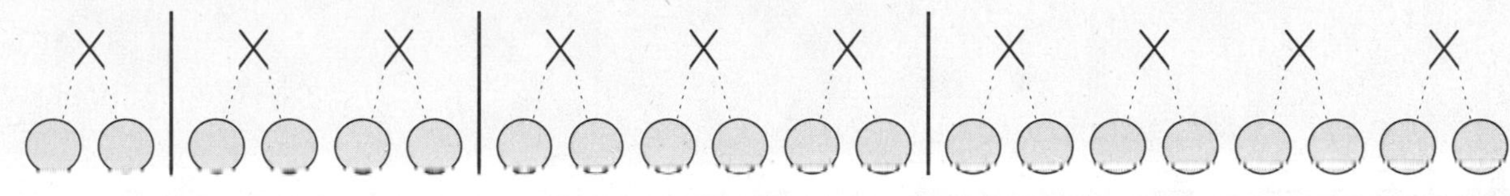

$$\frac{\text{Xs}}{\text{circles}} = \frac{1}{2} = \frac{2}{4} = \frac{3}{6} = \frac{4}{8}$$

$$\frac{1}{2} \times \frac{2}{2} = \frac{2}{4} \qquad \frac{1}{2} \times \frac{3}{3} = \frac{3}{6} \qquad \frac{1}{2} \times \frac{4}{4} = \frac{4}{8}$$

Complete each set of fractions to make the ratios equal.

1. $\frac{1}{3} = \frac{2}{6} = \frac{3}{\square} = \frac{\square}{12} = \frac{5}{\square} = \frac{\square}{18} = \frac{7}{\square} = \frac{8}{\square} = \frac{\square}{27}$

   $\frac{1}{3} \times \frac{2}{2} = \frac{2}{6}$

2. $\frac{5}{1} = \frac{10}{2} = \frac{\square}{3} = \frac{20}{\square} = \frac{\square}{5} = \frac{30}{\square} = \frac{\square}{7} = \frac{\square}{8} = \frac{45}{\square}$

3. $\frac{3}{1} = \frac{6}{2} = \frac{\square}{3} = \frac{12}{\square} = \frac{\square}{5} = \frac{18}{\square} = \frac{21}{\square} = \frac{\square}{8} = \frac{27}{\square}$

   $\frac{3}{1} \times \frac{}{3} = \frac{\square}{3}$

4. $\frac{1}{5} = \frac{2}{\square} = \frac{\square}{15} = \frac{4}{\square} = \frac{\square}{25} = \frac{6}{\square} = \frac{\square}{35} = \frac{8}{\square} = \frac{9}{\square}$

Equivalent Ratios

Fill in the $\square$ so that the ratios are equal.

1. $\frac{1}{2} = \frac{\square}{8}$

2. $\frac{5}{6} = \frac{15}{\square}$

3. $\frac{7}{8} = \frac{\square}{16}$

4. $\frac{1}{5} = \frac{\square}{45}$

5. $\frac{6}{7} = \frac{18}{\square}$

6. $\frac{3}{4} = \frac{\square}{20}$

7. $\frac{2}{9} = \frac{\square}{36}$

8. $\frac{4}{5} = \frac{\square}{25}$

9. $\frac{5}{\square} = \frac{1}{4}$

10. $\frac{18}{\square} = \frac{2}{5}$

11. $\frac{2}{\square} = \frac{1}{7}$

12. $\frac{\square}{49} = \frac{3}{7}$

13. $\frac{\square}{24} = \frac{3}{8}$

14. $\frac{5}{6} = \frac{\square}{12}$

15. $\frac{3}{4} = \frac{\square}{36}$

16. $\frac{5}{3} = \frac{10}{\square}$

17. $\frac{5}{4} = \frac{\square}{16}$

18. $\frac{\square}{64} = \frac{5}{8}$

19. $\frac{5}{7} = \frac{\square}{42}$

20. $\frac{3}{7} = \frac{21}{\square}$

21. $\frac{10}{\square} = \frac{2}{5}$

22. $\frac{1}{8} = \frac{5}{\square}$

23. $\frac{\square}{27} = \frac{1}{9}$

24. $\frac{3}{2} = \frac{\square}{4}$

Find a Pattern

A series of ratios form a pattern. Find the pattern and complete each table.

1. For every one can, there are three tennis balls.

Cans	1	2	3	4	5	6	7	8
Tennis Balls	3	6						

Think: $\frac{1 \times 2}{3 \times 2} = \frac{2}{6}$

2. For every one pack of gum, there are five sticks.

Packs of Gum	1	3	5	7	9	11	13	15
Sticks of Gum	5				45			

Think: $\frac{1 \times \square}{5 \times \square} = \frac{9}{45}$

3. For every one touchdown, a team gets six points.

Touchdowns	1	7	3	2	9	4	6	5
Points				12				30

Think: $\frac{1 \times 2}{\square \times 2} = \frac{2}{12}$

4. For every one dollar, you could get four quarters.

Dollars	1	3	7	2	10	6	9	4
Quarters		12	28					

Think: $\frac{1 \times \square}{\square \times \square} = \frac{3}{12}$

Fill in the Table

Complete the tables to show the amount paid for the number of hours worked.

1.

Hours Worked	1	2	3	8	40
Paid	$9		$27		

Think: $\frac{1}{9} \times \frac{\square}{\square} = \frac{3}{27}$

2.

Hours Worked	1	2	3	4	5
Paid	$8.75			$35	

Think: $\frac{1}{8.75} \times \frac{\square}{\square} = \frac{4}{35}$

3.

Hours Worked	1	8	10	40	80
Paid	$6.25				

4.

Hours Worked	1	4	8	20	40
Paid	$12.95				

From Words to Ratios

Ratios can be used to solve many problems. The first step is to write a statement as a ratio.

	Write a Ratio	Simplify
Sam worked 3 hours and earned $18.	$\frac{3}{18}$ ← hours / ← dollars	$\frac{3 \div 3}{18 \div 3} = \frac{1}{6}$

Write each ratio in its simplest form.

1. 8 miles in 2 hours $\frac{8}{2} = \frac{\square}{\square}$

2. 3 eggs are needed to make 12 cupcakes. $\frac{\square}{\square} = \frac{\square}{\square}$

3. $24 for working 3 hours $\frac{\square}{\square} = \frac{\square}{\square}$

4. 4 gallons of paint for $36 $\frac{\square}{\square} = \frac{\square}{\square}$

5. 100 miles on 5 gallons $\frac{\square}{\square} = \frac{\square}{\square}$

6. The box is 6 inches wide and 10 inches long. $\frac{\square}{\square} = \frac{\square}{\square}$

7. 5 tickets cost $35. $\frac{\square}{\square} = \frac{\square}{\square}$

8. Sue bicycled 27 miles in 3 hours. $\frac{\square}{\square} = \frac{\square}{\square}$

Unit Rates

A unit rate shows a ratio with a denominator of 1.

> *Per* means each or one.

60 miles per hour means 60 miles each hour: $\frac{60}{1}$

32 feet per second means 32 feet each second: $\frac{32}{1}$

$3.25 per square yard means $3.25 for one square yard: $\frac{\$3.25}{1}$

Write each ratio in fraction form. Circle the word that tells you there is a denominator of 1.

1. $\frac{\square}{1}$ 50 miles (per) hour

2. $\frac{\square}{\square}$ 3 tennis balls per can

3. $\frac{\square}{\square}$ 9 revolutions each minute

4. $\frac{\square}{\square}$ 50 tablets in one bottle

5. $\frac{\square}{\square}$ 25 miles per gallon

Write as a Ratio in Fraction Form

Write a fraction to show how the first number compares with the second. If there is a word that tells you the denominator is 1, circle the word.

1. $\frac{46}{1}$ 46 miles (per) hour
2. — Sue saves $48 in 5 weeks.
3. — 7 compared to 20
4. — 11 hits in 9 games
5. — 1 out of 5
6. — $7 per student
7. — 1 ounce equals 28 grams.
8. — 8 pounds for $9.85
9. — 32 feet per second
10. — 9 : 11
11. — $25 for each shirt
12. — 55 miles per hour
13. — 5 items for $16
14. — 36 miles per gallon
15. — $2 per pound
16. — 129 points in 8 games
17. — $9 per hour
18. — $5.00 per dozen
19. — 17 students per teacher
20. — $25 for working 3 hours

Writing Unit Rates

To find the unit rate of a ratio, simplify the fraction. The denominator must be 1.

Find the unit rate for each problem.

1. 85 miles on 5 gallons

$\frac{85}{5} \div \frac{5}{5} = \frac{17}{1}$ ← a denominator of one

Unit rate: 17 miles per gallon

2. 165 miles in 3 hours

$\frac{165}{3} \div \frac{3}{3} = \frac{\square}{1}$

Unit rate: ______ miles per hour

3. 64 ounces in 4 cans

$\frac{64}{4} = \frac{\square}{1}$

Unit rate: ______ ounces per can

4. 56 apples in 7 boxes

$\frac{\square}{\square} = \frac{\square}{1}$

Unit rate: ______ apples per box

5. 63 ounces in 9 cups

$\frac{\square}{\square} = \frac{\square}{1}$

Unit rate: ______ ounces per cup

6. $21 for 3 tickets

$\frac{\square}{\square} = \frac{\square}{1}$

Unit rate: ______ per ticket

7. 360 miles in 8 hours

$\frac{\square}{\square} = \frac{\square}{1}$

Unit rate: ______ miles per hour

8. 120 miles on 5 gallons

$\frac{\square}{\square} = \frac{\square}{1}$

Unit rate: ______ miles per gallon

Comparing Unit Rates

Number Relation Symbols
$<$ is less than
$>$ is greater than
$=$ is equal to

Write each statement as a ratio. Simplify each ratio to find the unit rate. Then compare the unit rates using the symbols above.

Fill in the symbol after you compare the unit rates.

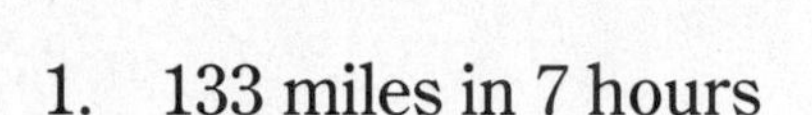

1. 133 miles in 7 hours ______ 108 miles in 6 hours
   (<, >, or =)

   $\frac{133}{7} = \frac{19}{1}$ $\qquad$ $\frac{108}{6} = \frac{\square}{1}$

   Unit rate: 19 miles per hour $\qquad$ Unit rate: ______ miles per hour

2. 72 miles on 3 gallons ______ 120 miles on 5 gallons
   (<, >, or =)

   $\frac{72}{3} = \frac{\square}{1}$ $\qquad$ $\frac{\square}{\square} = \frac{\square}{1}$

   Unit rate: ______ miles per gallon $\qquad$ Unit rate: ______ miles per gallon

3. $60 for 4 tickets ______ $90 for 5 tickets
   (<, >, or =)

   $\frac{\square}{\square} = \frac{\square}{1}$ $\qquad$ $\frac{\square}{\square} = \frac{\square}{1}$

   Unit rate: $______ for one ticket $\qquad$ Unit rate: $______ for one ticket

4. 425 meters in 5 hours ______ 294 meters in 3 hours
   (<, >, or =)

   $\frac{\square}{\square} = \frac{\square}{1}$ $\qquad$ $\frac{\square}{\square} = \frac{\square}{1}$

   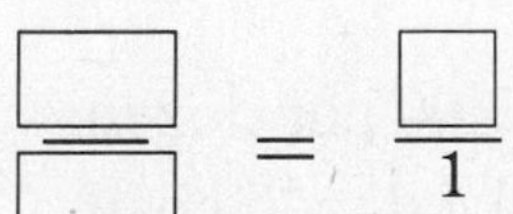

   Unit rate: ______ meters per hour $\qquad$ Unit rate: ______ meters per hour.

Find the Rates

Rates are ratios that compare different units of measurement.

Doug earns $6 per hour for mowing lawns.
How much will he earn in 4 hours?

$6 per hour   6 : 1
means

You know that Doug earns $6 in 1 hour.
To find how much he earns in 4 hours, multiply.

$$\$6 \times 4 = \$24$$

Use your understanding of ratios to find the amounts below.

1. Jack earns $8 per hour.
   a) How much money would he earn in 3 hours? ______________
   b) How much money would he earn in 8 hours? ______________

2. If Tim drives 45 miles per hour,
   a) how many miles could he drive in 8 hours? ______________
   b) how many miles could he drive in 4 hours? ______________

3. Jan's car gets 23 miles per gallon.
   a) How many miles can she drive on 15 gallons? ______________
   b) How many miles can she drive on 22 gallons? ______________

4. Pork chops cost $1.39 per pound.
   a) How much will 3 pounds of pork chops cost? ______________
   b) How much will 5 pounds of pork chops cost? ______________

5. The material costs $8.25 per yard.
   a) How much will 3 yards of material cost? ______________
   b) How much will 7 yards of material cost? ______________

6. Taffy apples cost $5.00 per dozen.
   a) How much will 4 dozen cost? ______________
   b) How much will 9 dozen cost? ______________

Find the Cost

Apples
$.99 per pound

Film
$1.45 per box

Cola
$4.25 per pack

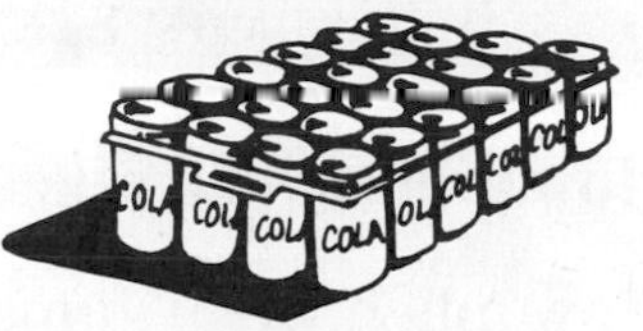

Milk
$.78 per quart

Gasoline
$1.19 per gallon

Candy
$2.25 per sack

Use the rates from the pictures to complete the tables.

1.

Item	Cost
1 pound apples	a)
2 pounds apples	b)
3 pounds apples	c)
4 pounds apples	d)

3.

Item	Cost
1 gallon gasoline	a)
2 gallons gasoline	b)
11 gallons gasoline	c)
15 gallons gasoline	d)

5.

Item	Cost
1 sack candy	a)
2 sacks candy	b)
5 sacks candy	c)
7 sacks candy	d)

2.

Item	Cost
1 box film	a)
4 boxes film	b)
6 boxes film	c)
10 boxes film	d)

4.

Item	Cost
1 quart milk	a)
2 quarts milk	b)
4 quarts milk	c)
5 quarts milk	d)

6.

Item	Cost
1 pack cola	a)
3 packs cola	b)
4 packs cola	c)
6 packs cola	d)

Ratios as Rates

Use the information given to complete each problem.

3 items cost $12

1. a) How much will 6 items cost? ____
   b) How much will 9 items cost? ____
   c) How much will 12 items cost? ____

$5 for 3 pairs of socks

2. a) 6 pairs will cost ____
   b) 9 pairs will cost ____
   c) 12 pairs will cost ____

4 apples cost $.92

3. a) How much will 8 apples cost? ____
   b) How much will 12 apples cost? ____
   c) How much will 16 apples cost? ____

$4.50 for 2 pens

4. a) How much will 4 cost? ____
   b) How much will 6 cost? ____
   c) How much will 8 cost? ____

3 tapes for $8

5. a) How much will 9 cost? ____
   b) How much will 15 cost? ____
   c) How much will 21 cost? ____

2 candy bars for $1.35

6. a) How much will 6 cost? ____
   b) How much will 10 cost? ____
   c) How much will 16 cost? ____

$4.25 for 3 ice cream cones

7. a) 9 will cost $ ____
   b) 15 will cost $ ____
   c) 27 will cost $ ____

5 tablets for $1.50

8. a) How much will 15 tablets cost? ____
   b) How much will 25 tablets cost? ____
   c) How much will 40 tablets cost? ____

Measurement Ratios

Sometimes you will need to compare measurements that are in different units (inches, feet, yards, etc.).

- First, write the measurements in the same units.
- Then, write a fraction ratio to compare the measurements.

12 inches = 1 foot 3 feet = 1 yard

Compare				Simplified Ratio
6 inches to 1 foot	=	$\frac{6 \text{ inches}}{12 \text{ inches}}$	=	$\frac{1}{2}$

Compare the following measurements as fraction ratios.

	Compare				Simplified Ratio
1.	3 inches to 1 foot	=	$\frac{3 \text{ inches}}{12 \text{ inches}}$ (inches in 1 foot)	=	$\frac{\square}{\square}$
2.	12 inches to 1 foot	=	$\frac{\square \text{ inches}}{\square \text{ inches}}$	=	$\frac{\square}{\square}$
3.	36 inches to 1 foot	=	$\frac{\square \text{ inches}}{\square \text{ inches}}$	=	$\frac{\square}{\square}$
4.	6 feet to 1 yard	=	$\frac{6 \text{ feet}}{3 \text{ feet}}$ (feet in 1 yard)	=	$\frac{\square}{\square}$
5.	3 feet to 1 yard	=	$\frac{\square \text{ feet}}{\square \text{ feet}}$	=	$\frac{\square}{\square}$
6.	1 foot to 1 yard	=	$\frac{\square \text{ foot}}{\square \text{ feet}}$	=	$\frac{\square}{\square}$

Comparing Inches and Feet

Measurement ratios may be compared in the same units. To do so, change the larger units to smaller units.

12 inches = 1 foot

Compare		Simplified Ratio
3 inches to 1 foot	$= \frac{3 \text{ inches}}{12 \text{ inches}}$	$= \frac{1}{4}$

Change the feet to inches and write a ratio to compare.

Compare		Simplified Ratio
1. 4 inches to 2 feet	$= \frac{4 \text{ inches}}{24 \text{ inches}}$ (24: inches in 2 feet)	$= \frac{\square}{\square}$
2. 3 inches to 1 foot	$= \frac{\square \text{ inches}}{12 \text{ inches}}$ (12: inches in 1 foot)	$= \frac{\square}{\square}$
3. 10 inches to 5 feet	$= \frac{\square \text{ inches}}{60 \text{ inches}}$ (60: inches in 5 feet)	$= \frac{\square}{\square}$
4. 12 inches to 3 feet	$= \frac{\square \text{ inches}}{\square \text{ inches}}$	$= \frac{\square}{\square}$
5. 9 inches to 1 foot	$= \frac{\square \text{ inches}}{\square \text{ inches}}$	$= \frac{\square}{\square}$

Comparing Feet and Yards

Measurement ratios may be compared in the same units. To do so, change the larger units to smaller units.

	Compare		Simplified Ratio
3 feet = 1 yard	2 feet to 1 yard	$= \frac{2 \text{ feet}}{3 \text{ feet}}$	$= \frac{2}{3}$

Change the yards to feet and write a ratio to compare.

	Compare			Simplified Ratio
1.	4 feet to 4 yards	$= \frac{\square \text{ feet}}{12 \text{ feet}}$ (feet in 4 yards)	=	$\frac{\square}{\square}$
2.	3 feet to 2 yards	$= \frac{\square \text{ feet}}{6 \text{ feet}}$ (feet in 2 yards)	=	$\frac{\square}{\square}$
3.	6 feet to 5 yards	$= \frac{\square \text{ feet}}{\square \text{ feet}}$	=	$\frac{\square}{\square}$
4.	5 feet to 10 yards	$= \frac{\square \text{ feet}}{\square \text{ feet}}$	=	$\frac{\square}{\square}$
5.	2 feet to 6 yards	$= \frac{\square \text{ feet}}{\square \text{ feet}}$	=	$\frac{\square}{\square}$

Comparing Ounces and Pounds

Measurement ratios may be compared in the same units. To do so, change the larger units to smaller units.

16 ounces = 1 pound

Compare				Simplified Ratio
2 ounces to 1 pound	=	$\frac{2 \text{ ounces}}{16 \text{ ounces}}$	=	$\frac{1}{8}$

Change the pounds to ounces and write a ratio to compare.

	Compare				Simplified Ratio
1.	4 ounces to 1 pound	=	$\frac{\square \text{ ounces}}{16 \text{ ounces}}$ (16: ounces in 1 pound)	=	$\frac{\square}{\square}$
2.	6 ounces to 3 pounds	=	$\frac{\square \text{ ounces}}{48 \text{ ounces}}$ (48: ounces in 3 pounds)	=	$\frac{\square}{\square}$
3.	8 ounces to 2 pounds	=	$\frac{\square \text{ ounces}}{\square \text{ ounces}}$ (denominator: ounces in 2 pounds)	=	$\frac{\square}{\square}$
4.	2 ounces to 2 pounds	=	$\frac{\square \text{ ounces}}{\square \text{ ounces}}$	=	$\frac{\square}{\square}$
5.	10 ounces to 5 pounds	=	$\frac{\square \text{ ounces}}{\square \text{ ounces}}$	=	$\frac{\square}{\square}$

Money and Time Ratios

Measurement ratios may be compared in the same units. To do so, change the larger units to smaller units.

	Compare		Simplified Ratio
60 seconds = 1 minute	5 seconds to 1 minute	$= \frac{5 \text{ seconds}}{60 \text{ seconds}}$	$= \frac{1}{12}$

Change the larger unit to the smaller unit and write a ratio to compare.

	Compare		Simplified Ratio
1.	6 pennies to 2 dimes	$= \frac{\square \text{ pennies}}{20 \text{ pennies}}$ (pennies in 2 dimes)	$= \frac{\square}{\square}$
2.	2 nickels to 1 dollar	$= \frac{\square \text{ nickels}}{\square \text{ nickels}}$ (nickels in 1 dollar)	$= \frac{\square}{\square}$
3.	2 quarters to 2 dollars	$= \frac{\square \text{ quarters}}{\square \text{ quarters}}$	$= \frac{\square}{\square}$
4.	15 minutes to 1 hour	$= \frac{\square \text{ minutes}}{60 \text{ minutes}}$ (minutes in 1 hour)	$= \frac{\square}{\square}$
5.	1 minute to 100 seconds	(seconds in 1 minute →) $= \frac{\square \text{ seconds}}{\square \text{ seconds}}$	$= \frac{\square}{\square}$
6.	50 seconds to 2 minutes	$= \frac{\square \text{ seconds}}{\square \text{ seconds}}$	$= \frac{\square}{\square}$

Ratio Review

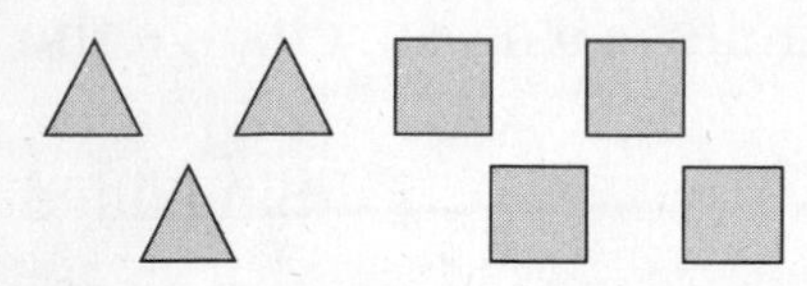

1. Compare the number of squares to triangles.

   a) ____ to ____ b) ____ : ____ c) $\frac{\square}{\square}$

Write each ratio as a fraction and simplify.

2. 13 : 26 $= \frac{\square}{\square} = \frac{\square}{\square}$

3. 48 to 6 $= \frac{\square}{\square} = \frac{\square}{\square}$

4. Make the ratios equal.

   $\frac{3}{8} = \frac{12}{\square}$

Write each ratio in fraction form. Simplify when necessary.

5. A box is 8 inches wide and 12 inches long.

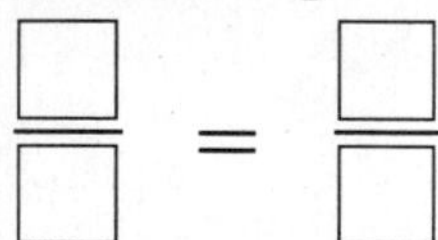

6. $45 for 3 shirts

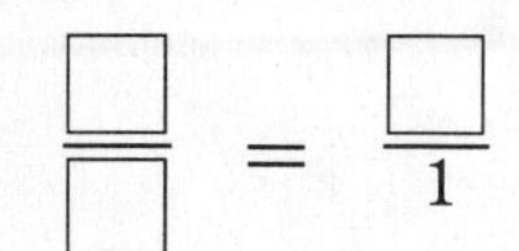

7. Compare the unit rates

   $25 for 5 tickets ________ $49 for 7 tickets

   <, >, or =

   $\frac{\square}{\square} = \frac{\square}{1}$

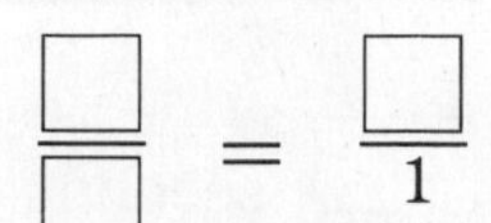

   Unit rate: $______ for one ticket Unit rate: $______ for one ticket

Use the information given to complete each problem.

3 tomatoes cost $.81

8. a) How much will 5 tomatoes cost? ____

   b) How much will 8 tomatoes cost? ____

   c) How much will 10 tomatoes cost? ____

Write a simplified ratio for each problem.

9. 15 inches to 2 feet $= \frac{\square}{\square} = \frac{\square}{\square}$

10. 6 feet to 4 yards $= \frac{\square}{\square} = \frac{\square}{\square}$

11. 12 ounces to 1 pound $= \frac{\square}{\square} = \frac{\square}{\square}$

12. 10 nickels to 2 dollars $= \frac{\square}{\square} = \frac{\square}{\square}$

13. 5 minutes to 1 hour $= \frac{\square}{\square} = \frac{\square}{\square}$

Ratio Applications

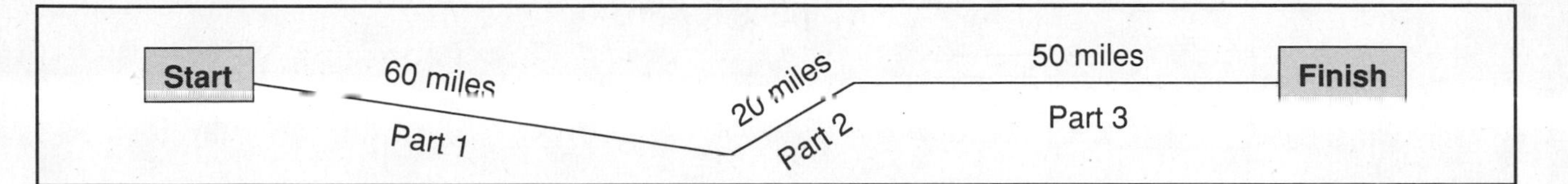

Use the map above to answer questions 1 through 5.

Simplified

1. What is the ratio of Part 1 to the whole trip? ____ : ____ $= \frac{\square}{\square} = \frac{\square}{\square}$

2. What is the ratio of the whole trip to Part 2? ____ : ____ $= \frac{\square}{\square} = \frac{\square}{\square}$

3. What is the ratio of Part 3 to the whole trip? ____ : ____ $= \frac{\square}{\square} = \frac{\square}{\square}$

4. What is the ratio of the whole trip to Part 1? ____ : ____ $= \frac{\square}{\square} = \frac{\square}{\square}$

5. What is the ratio of Parts 1 and 3 to the whole trip? ____ : ____ $= \frac{\square}{\square} = \frac{\square}{\square}$

A recipe for punch calls for 3 pints of Hawaiian Punch, 4 pints of orange juice, and 5 pints of ginger ale.

6. What is the ratio of ginger ale to the whole drink? ____ : ____ $= \frac{\square}{\square}$

7. What is the ratio of orange juice to the whole drink? ____ : ____ $= \frac{\square}{\square} = \frac{\square}{\square}$

8. What is the ratio of the whole drink to orange juice? ____ : ____ $= \frac{\square}{\square} = \frac{\square}{\square}$

Using Ratios

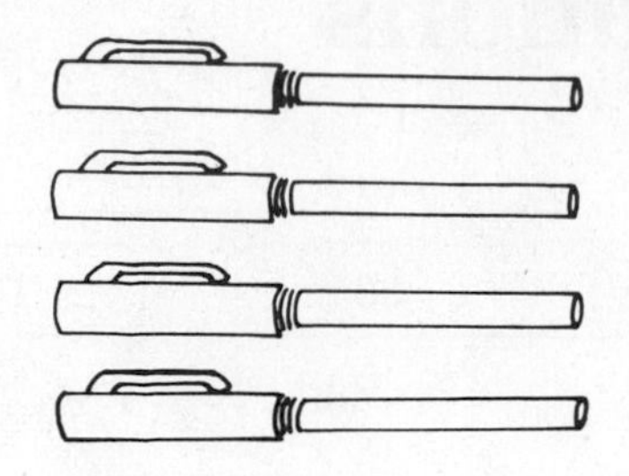

4 pens for \$6.00

How much do 8 pens cost?

Step 1: Write a ratio. 8 : 4

Step 2: Write as a fraction and simplify. $\frac{8}{4} = 2$ sets

Step 3: $\$6.00 \times 2 = \$12.00$

How much do 6 pens cost?

Step 1: Write a ratio. 6 : 4

Step 2: Write as a fraction and simplify. $\frac{6}{4} = \frac{3}{2}$ $(1\frac{1}{2})$ sets

Step 3: $\overset{3.00}{\cancel{\$6.00}} \times \frac{3}{\underset{1}{\cancel{2}}} = \$9.00$

Use a ratio to solve the problems.

SPECIAL!
2 for \$3.00

1. a) How much will 4 pints of ice cream cost?

   b) How much will 3 pints of ice cream cost?

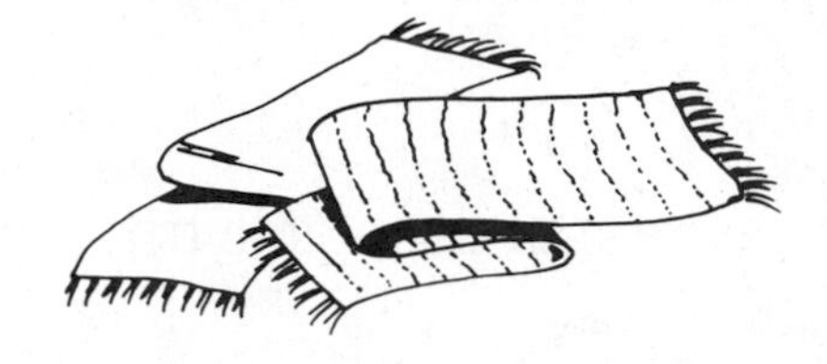

Scarves
2 for \$15.00

3. a) How much will 6 scarves cost?

   b) How much will 5 scarves cost?

3 for \$2.00

2. a) How much will 9 rolls of paper towels cost?

   b) How much will 12 rolls of paper towels cost?

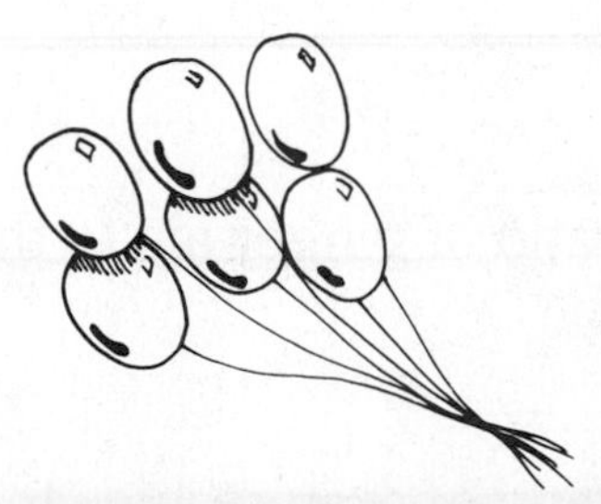

6 balloons
for \$10.00

4. a) How much will 24 balloons cost?

   b) How much will 9 balloons cost?

Real-Life Ratios

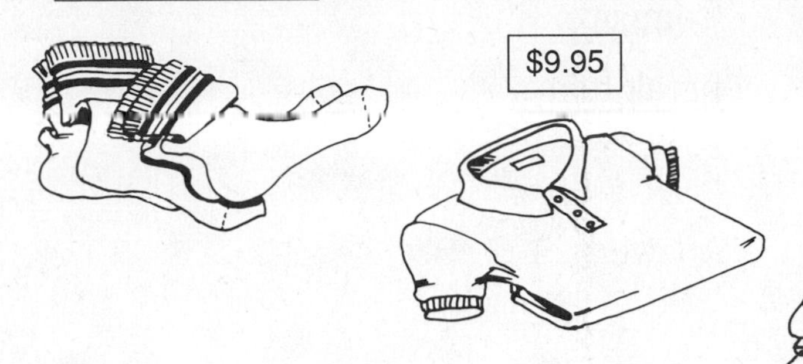

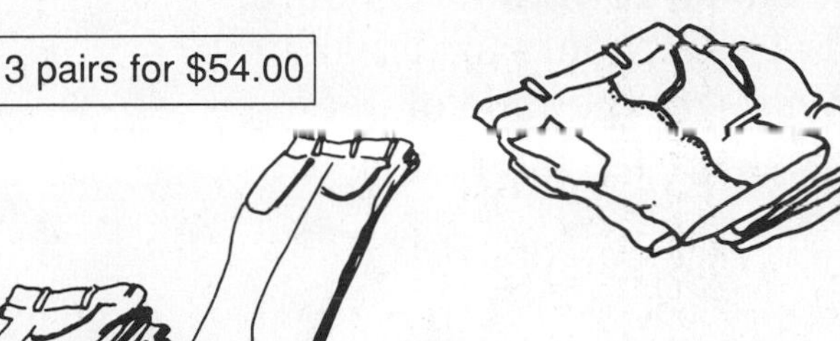

Use the pictures to solve the problems.

1. How much will it cost for:
   a) One pair of slacks? ______
   b) One pair of socks? ______
   c) One pair of shorts? ______

2. How much will it cost for:
   a) 6 pairs of shorts? ______
   b) 3 sport shirts? ______
   c) 8 pairs of socks? ______

3. If you want to buy 2 sport shirts, one pair of slacks, and 2 pairs of shorts, how much change will you get back from $70.00? ______

4. How much will it cost for:
   a) 4 pairs of socks? ______
   b) 3 pairs of shorts? ______
   c) 4 pairs of slacks? ______

5. How much will it cost for:
   a) 3 sport shirts? ______
   b) 3 pairs of socks? ______
   c) 2 pairs of slacks? ______

6. How much change will you get back from $15.00 if you buy one pair of shorts? ______

Seeing Ratios in Word Problems

Some word problems contain a ratio. These problems ask you to make a comparison.

Sam was at work for 9 hours. He took 2 hours for breaks. What was the ratio of his breaks to his hours at work?

a) 2 : 7　　c) 2 : 9

b) 7 : 2　　d) 9 : 2

Compare:

$$\frac{\text{break hours}}{\text{hours at work}} = \frac{2}{9}$$

Answer: c) 2 : 9

Circle the correct ratio for the problems below.

1. Scott repaired window sills for 5 hours. Out of 18 window sills, he repaired 15. What was the ratio between window sills fixed and hours worked?

   a) 5 : 15　　c) 18 : 15

   b) 5 : 18　　d) 15 : 5

2. Janet's band played for 3 hours. They took two 30-minute breaks. What was the ratio between the length of the band's breaks and the hours played?

   a) 2 : 30　　c) 30 : 3

   b) 3 : 30　　d) 1 : 3

3. In 7 days, Mary's Restaurant sold 74 dinner specials. The restaurant also sold 121 regular dinners. What was the ratio between dinner specials and regular dinners sold?

   a) 7 : 74　　c) 121 : 7

   b) 121 : 74　　d) 74 : 121

4. At basketball practice, Kathy shot 100 free throws. She made 82 shots, and missed 18. What was the ratio between the shots she missed and the shots she attempted?

   a) 82 : 100　　c) 18 : 82

   b) 18 : 100　　d) 100 : 18

Ratio Relationships

1. What is the ratio of shaded squares to all squares? ____ : ____

2. What ratio represents the statement "seven out of nine square regions"? ____ : ____

3. If 6 pounds cost $5.40, what is the cost per pound? ______

4. A pinch hitter was up at bat 4 times and got 3 hits. Compare the number of hits to the number of times at bat. ____ : ____

5. The ratio of whales to dolphins is 9 to 3. How many whales are there per dolphin?
   ____ : ____ = ____ : ____ (simplified)

6. An item originally priced at $20.00 is reduced to $15.00.
   a) What is the ratio of the reduced price to the original price?
      ____ : ____ = ____ : ____ (simplified)
   b) What is the ratio of the amount saved to the original price?
      ____ : ____ = ____ : ____ (simplified)

7. If $\frac{1}{4}$ inch on a scale drawing represents 5 miles, what does 1 (one) inch represent? ______

8. Lana earns $6.95 per hour. How much does she earn in 8 hours? ______

What Is a Proportion?

You can use a proportion to solve many different types of problems.

A **proportion** is made up of two equal ratios.

You can multiply or divide to find equal ratios.

Multiply by a value of one.

$$\frac{2}{3}\begin{matrix}\times\\\times\end{matrix}\frac{2}{2} = \frac{4}{6}$$

So, $\frac{2}{3} = \frac{4}{6}$ is a proportion.

Divide by a value of one.

$$\frac{6}{9}\begin{matrix}\div\\\div\end{matrix}\frac{3}{3} = \frac{2}{3}$$

So, $\frac{6}{9} = \frac{2}{3}$ is a proportion.

Solve each proportion by finding the missing term.

1. $\frac{3 \times 2}{4 \times 2} = \frac{\square}{8}$

2. $\frac{15 \div 5}{5 \div 5} = \frac{\square}{1}$

3. $\frac{1}{4} = \frac{3}{\square}$

4. $\frac{3}{2} = \frac{15}{\square}$

5. $\frac{4 \times 4}{7 \times 4} = \frac{\square}{28}$

6. $\frac{\square}{6} = \frac{10}{12}$

7. $\frac{3}{\square} = \frac{15}{20}$

8. $\frac{\square}{10} = \frac{1}{2}$

9. $\frac{6 \quad \div 3}{\square \quad \div 3} = \frac{2}{9}$

10. $\frac{40}{100} = \frac{\square}{10}$

11. $\frac{9}{7} = \frac{18}{\square}$

12. $\frac{\square}{100} = \frac{3}{4}$

Simplify One Ratio

To find a missing term in a proportion, you may want to simplify one of the ratios.

EXAMPLE

Find the missing term.

$$\frac{3}{6} = \frac{\square}{8}$$

STEP 1

Simplify $\frac{3}{6}$ to $\frac{1}{2}$.

$$\frac{\cancel{3}^{1}}{\cancel{6}_{2}} = \frac{\square}{8}$$

STEP 2

Solve for the missing term.

$$\frac{1 \times}{2 \times} \frac{4}{4} = \frac{\boxed{4}}{8}$$

Think: What number do you multiply 2 by to get 8?

Simplify if necessary. Find each missing term.

1. $\frac{\cancel{4}^{1} \times 5}{\cancel{8}_{2} \times 5} = \frac{\square}{10}$

2. $\frac{6}{15} = \frac{8}{\square}$

3. $\frac{\square}{18} = \frac{5}{30}$

4. $\frac{8}{\square} = \frac{10}{15}$

5. $\frac{\square}{12} = \frac{\cancel{6}^{3}}{\cancel{8}_{4}}$

6. $\frac{\square}{24} = \frac{15}{20}$

7. $\frac{18}{\square} = \frac{30}{5}$

8. $\frac{15}{9} = \frac{\square}{12}$

9. $\frac{\cancel{7}^{1}}{\cancel{21}_{3}} = \frac{9}{\square}$

10. $\frac{10}{14} = \frac{\square}{21}$

11. $\frac{6}{2} = \frac{15}{\square}$

12. $\frac{30}{25} = \frac{\square}{15}$

Read the Proportion

A proportion is a statement that two ratios are equal.

Proportion

$$\frac{3}{4} = \frac{9}{12}$$

Read: "3 is to 4 as 9 is to 12."

Write the proportions as fraction ratios.

Example: 2 is to 3 as 6 is to 9 $\frac{2}{3} = \frac{6}{9}$

1. 1 is to 4 as 2 is to 8 $\frac{1}{4} = \frac{\square}{\square}$

2. 7 is to 3 as 14 is to 6 $\frac{\square}{\square} = \frac{\square}{\square}$

3. 2 is to 5 as 4 is to 10 $\frac{\square}{\square} = \frac{\square}{\square}$

4. 5 is to 8 as 10 is to 16 $\frac{\square}{\square} = \frac{\square}{\square}$

5. 8 is to 12 as 4 is to 6 $\frac{\square}{\square} = \frac{\square}{\square}$

6. 50 is to 25 as 2 is to 1 $\frac{\square}{\square} = \frac{\square}{\square}$

Two Equal Ratios

Two ratios are equal when their **cross products** are equal.

The cross products are: 2×6 and 4×3.

$$\frac{2}{3} = \frac{4}{6}$$

$$2 \times 6 = 4 \times 3$$
$$12 = 12$$

The cross products are: 1×9 and 3×3.

$$\frac{1}{3} = \frac{3}{9}$$

$$1 \times 9 = 3 \times 3$$
$$9 = 9$$

Find the cross products for each proportion.

1. $\frac{2}{3} = \frac{6}{9}$

   $2 \times 9 = 6 \times 3$

   $18 =$ ______

2. $\frac{7}{3} = \frac{21}{9}$

   ___ × ___ = ___ × ___

   ___ = ___

3. $\frac{4}{5} = \frac{16}{20}$

   ___ × ___ = ___ × ___

   ___ = ___

4. $\frac{8}{5} = \frac{40}{25}$

   ___ × ___ = ___ × ___

   ___ = ___

5. $\frac{6}{18} = \frac{12}{36}$

   ___ × ___ = ___ × ___

   ___ = ___

6. $\frac{5}{1} = \frac{10}{2}$

   ___ × ___ = ___ × ___

   ___ = ___

7. $\frac{8}{3} = \frac{16}{6}$

   ___ × ___ = ___ × ___

   ___ = ___

8. $\frac{7}{8} = \frac{91}{104}$

   ___ × ___ = ___ × ___

   ___ = ___

9. $\frac{10}{15} = \frac{30}{45}$

   ___ × ___ = ___ × ___

   ___ = ___

Cross Products

Two ratios are equal when their cross products are equal.

$\frac{2}{3} \times \frac{4}{6}$

$2 \times 6 = 4 \times 3$
$12 = 12$

The cross products are equal:

So, $\frac{2}{3} = \frac{4}{6}$

$\frac{2}{5} \times \frac{3}{4}$

$2 \times 4 \neq 3 \times 5$
$8 \neq 15$

The cross products are not equal:

So, $\frac{2}{5} \neq \frac{3}{4}$

- Use cross products to determine whether or not the ratios are equal.
- Use the symbols = (equal to) or ≠ (not equal to) in the  .

1. $\frac{1}{4} \times \frac{3}{12}$

   $1 \times 12 \quad 3 \times 4$

   12 (=) ___

2. $\frac{5}{6} \times \frac{2}{3}$

   $5 \times 3 \quad 2 \times 6$

   ___ (≠) ___

3. $\frac{21}{28} \times \frac{3}{4}$

   ___ × ___ ___ × ___

   ___ ◯ ___

4. $\frac{42}{56} \times \frac{7}{9}$

   ___ × ___ ___ × ___

   ___ ◯ ___

5. $\frac{9}{13} \times \frac{63}{91}$

   ___ × ___ ___ × ___

   ___ ◯ ___

6. $\frac{7}{3} \times \frac{14}{6}$

   ___ × ___ ___ × ___

   ___ ◯ ___

7. $\frac{5}{12} \times \frac{80}{192}$

   ___ × ___ ___ × ___

   ___ ◯ ___

8. $\frac{4}{7} \times \frac{52}{91}$

   ___ × ___ ___ × ___

   ___ ◯ ___

9. $\frac{12}{15} \times \frac{24}{40}$

   ___ × ___ ___ × ___

   ___ ◯ ___

Proportion Readiness

If the cross products are equal, then the ratios are equal. Use cross products to see if the two ratios form a proportion.

Find the cross products. Write = (equal to) or ≠ (not equal to) in the  .

1. $\frac{1}{2}$ (=) $\frac{5}{10}$
   [10] [10]
   1 × 10 5 × 2

2. $\frac{3}{5}$ (≠) $\frac{2}{3}$
   [9] [10]
   3 × 3 2 × 5

3. $\frac{4}{6}$ ◯ $\frac{12}{18}$

4. $\frac{9}{8}$ ◯ $\frac{54}{48}$

5. $\frac{6}{8}$ ◯ $\frac{9}{16}$

6. $\frac{12}{3}$ ◯ $\frac{36}{9}$

7. $\frac{8}{4}$ ◯ $\frac{32}{16}$

8. $\frac{11}{13}$ ◯ $\frac{8}{12}$

9. $\frac{8}{24}$ ◯ $\frac{3}{9}$

10. $\frac{5}{8}$ ◯ $\frac{45}{72}$

11. $\frac{18}{14}$ ◯ $\frac{9}{7}$

12. $\frac{35}{4}$ ◯ $\frac{27}{3}$

Find the Unknown Term

You can use cross products to find a missing number in a proportion. Use an n to stand for the missing number.

A missing number in a proportion is called a **term.**

To solve a proportion: $\frac{6}{27} = \frac{2}{n}$

Step 1: Write the cross products. $6 \times n = 2 \times 27$

Step 2: Multiply the two numbers in the cross product. $6 \times n = 54$

Step 3: Divide by the third number to find what n is. $n = 54 \div 6$

Step 4: Solve for n. $n = 9$

Find the missing terms in the proportions below.

1. $\frac{3}{4} = \frac{n}{32}$

   $3 \times 32 = n \times 4$

   $96 = n \times 4$

   $96 \div 4 = n$

   ______ $= n$

4. $\frac{n}{9} = \frac{2}{6}$

   $n \times 6 = 2 \times 9$

   $n \times 6 = 18$

   $n = 18 \div 6$

   $n =$ ______

2. $\frac{8}{n} = \frac{2}{5}$

5. $\frac{n}{8} = \frac{15}{20}$

3. $\frac{n}{14} = \frac{3}{21}$

6. $\frac{n}{15} = \frac{4}{5}$

Solve and Check

In the proportion $\frac{6}{8} = \frac{n}{20}$, Sam got an answer of 15 for n and Jackie got an answer of 10 for n. Who was correct, Sam or Jackie?

		Sam's check	Jackie's check
Step 1:	Write the solution for n.	$\frac{6}{8} = \frac{15}{20}$	$\frac{6}{8} = \frac{10}{20}$
Step 2:	Cross multiply.	$6 \times 20 = 15 \times 8$	$6 \times 20 = 10 \times 8$
Step 3:	Compare.	$120 = 120$ ← correct	$120 \neq 80$ ← incorrect

Sam's answer is correct because the cross products are equal.

Solve for n. Check each answer by multiplying the cross products.

1. $\frac{4}{n} = \frac{3}{15}$

2. $\frac{9}{6} = \frac{15}{n}$

3. $\frac{28}{4} = \frac{n}{5}$

4. $\frac{n}{6} = \frac{6}{4}$

5. $\frac{12}{n} = \frac{15}{10}$

6. $\frac{10}{6} = \frac{n}{9}$

7. $\frac{15}{n} = \frac{2}{12}$

8. $\frac{12}{18} = \frac{2}{n}$

9. $\frac{14}{n} = \frac{21}{6}$

Apply Your Skills

Solve for n in the proportions below.

1. $\frac{1}{6} = \frac{n}{30}$

2. $\frac{9}{n} = \frac{6}{8}$

3. $\frac{n}{18} = \frac{5}{6}$

4. $\frac{25}{10} = \frac{15}{n}$

5. $\frac{7}{8} = \frac{n}{56}$

6. $\frac{n}{5} = \frac{18}{2}$

7. $\frac{4}{6} = \frac{10}{n}$

8. $\frac{100}{n} = \frac{5}{2}$

9. $\frac{5}{n} = \frac{1}{200}$

10. $\frac{4}{7} = \frac{16}{n}$

11. $\frac{3}{8} = \frac{n}{24}$

12. $\frac{8}{2} = \frac{12}{n}$

13. $\frac{n}{36} = \frac{8}{9}$

14. $\frac{2}{5} = \frac{n}{25}$

15. $\frac{2}{n} = \frac{3}{27}$

Proportions with Fractions

Find the missing terms in the proportions below. Change all answers to mixed numbers.

1. $\frac{2}{3} = \frac{n}{5}$

   $2 \times 5 = n \times 3$

   $10 = n \times 3$

   $10 \div 3 = n$

   $3\frac{1}{3} = n$

2. $\frac{n}{3} = \frac{8}{5}$

3. $\frac{3}{7} = \frac{5}{n}$

4. $\frac{6}{n} = \frac{4}{7}$

5. $\frac{8}{7} = \frac{9}{n}$

   $8 \times n = 9 \times 7$

   $8 \times n = 63$

   $n = 63 \div 8$

   $n =$ ______

6. $\frac{8}{7} = \frac{n}{8}$

7. $\frac{n}{6} = \frac{3}{5}$

8. $\frac{9}{n} = \frac{4}{7}$

Proportions with Decimals

Use the cross-product method to solve proportions with decimals in them.

1. $\frac{3}{4} = \frac{3.6}{n}$

$3 \times n = 3.6 \times 4$

$3 \times n = 14.4$

$n = 14.4 \div 3$

$n = 4.8$

5. $\frac{5}{2.5} = \frac{n}{17.5}$

$5 \times 17.5 = n \times 2.5$

$87.5 = n \times 2.5$

$87.5 \div 2.5 = n$

$\_\_\_\_\_ = n$

2. $\frac{n}{1.2} = \frac{6}{3.6}$

6. $\frac{.14}{.28} = \frac{3}{n}$

3. $\frac{7}{n} = \frac{5}{9.35}$

7. $\frac{8}{n} = \frac{13}{2.6}$

4. $\frac{n}{2} = \frac{3.54}{3}$

8. $\frac{6}{1.26} = \frac{n}{2.52}$

Missing Term

Write a proportion for each sentence and solve for the missing term.

Sentence	Proportion	Missing Term
1. 3 is to 4 as n is to 16	$\frac{3}{4} = \frac{n}{16}$	$n =$ ______ whole number
2. 4 is to 6 as 10 is to n	$\frac{\square}{\square} = \frac{\square}{\square}$	______ whole number
3. n is to 20 as 5 is to 8	$\frac{\square}{\square} = \frac{\square}{\square}$	______ mixed number
4. 2.5 is to n as 15 is to 18	$\frac{\square}{\square} = \frac{\square}{\square}$	______ whole number
5. 2.78 is to 2 as n is to 5	$\frac{\square}{\square} = \frac{\square}{\square}$	______ decimal
6. 8 is to 5 as 13 is to n	$\frac{\square}{\square} = \frac{\square}{\square}$	______ mixed number
7. 3.25 is to 1 as n is to 6	$\frac{\square}{\square} = \frac{\square}{\square}$	______ decimal
8. n is to 4 as 15 is to 6	$\frac{\square}{\square} = \frac{\square}{\square}$	______ whole number

Proportion Review

Find the missing term in each proportion. Simplify first if necessary.

1. 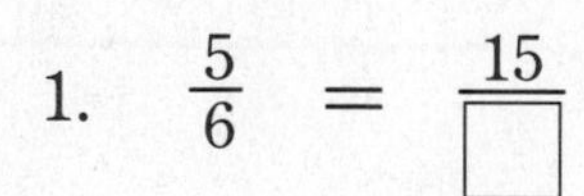

2. $\frac{4}{8} = \frac{\square}{12}$

3. Write the proportion using fraction ratios.

   9 is to 27 as 6 is to 18

   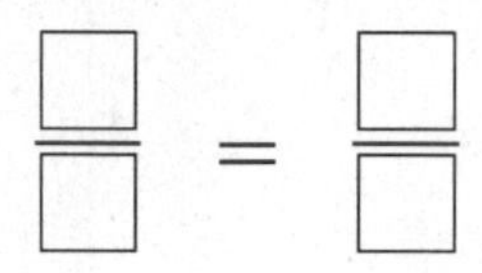

4. Find the cross products.

   $\frac{7}{3} = \frac{21}{9}$

   ___ × ___ = ___ × ___

   ___ = ___

5. Find the cross products and compare using = or ≠.

   $\frac{5}{8}$ ◯ $\frac{40}{63}$

   $\square$ $\square$

6. Find the missing term.

   $\frac{n}{30} = \frac{2}{5}$

   $n \times 5 = 2 \times 30$

   $n \times 5 = 60$

   $n = 60 \div$ ____

   $n =$ _____

7. Solve for n.

   $\frac{4}{3} = \frac{16}{n}$

8. Solve for n. Change the answer to a mixed number.

   $\frac{9}{n} = \frac{5}{6}$

9. Use the cross-product method to solve the proportion.

   $\frac{n}{3} = \frac{4.15}{2.49}$

Write a proportion and solve for n.

10. 7.25 is to 1 as n is to 6

11. n is to 4 as 15 is to 10

Proportions in Problem Solving

1. Read the problem carefully.
2. Represent the quantity you do not know by the letter n.
3. Write the units next to the numbers when writing the proportion.
4. Make sure the same units hold corresponding positions in the two ratios of the proportions.

Traveling at the rate of 110 miles in 2 hours, how far can you go in 5 hours?

The first ratio compares **miles** to **hours:** $\frac{110}{2}\ \frac{\text{miles}}{\text{hours}}$

The second ratio must also compare miles to hours: $\frac{n}{5}\ \frac{\text{miles}}{\text{hours}}$

The correct proportion is: $\frac{110}{2}\ \frac{\text{miles}}{\text{hours}} = \frac{n}{5}\ \frac{\text{miles}}{\text{hours}}$

Set up the proportion. **Do not solve.**

If 5 cans of soup cost $2.50, how much will 9 cans of soup cost?

1. The first ratio compares ___cans___ to ___cost___ $\frac{5}{\$2.50}\ \frac{\text{cans}}{\text{cost}}$

2. The second ratio must also compare ________ to ________ $\frac{9}{\square}\ \frac{\text{cans}}{\text{cost}}$

3. The correct proportion is: $\frac{\square}{\square}\ \frac{\text{cans}}{\text{cost}} = \frac{\square}{\square}\ \frac{\_\_\_\_}{\_\_\_\_}$

Setting Up Proportions

Make sure the same units hold corresponding positions in the two ratios of the proportion.

You can travel 28 miles on 2 gallons of gasoline. At the same rate, how far can you travel on 7 gallons of gasoline?

The first ratio compares miles to gallons.

$$\frac{28 \text{ miles}}{2 \text{ gallons}} = \frac{n \text{ miles}}{7 \text{ gallons}}$$

The second ratio must also compare miles to gallons.

Set up the proportions. Write in the numbers and the labels. **Do not solve.**

1. Kevin runs 3 miles in 24 minutes. At the same rate, how long does it take him to run 10 miles?

$$\frac{\square \text{ miles}}{\square \text{ minutes}} = \frac{\square \text{ miles}}{\square \text{ minutes}}$$

2. José drove 675 miles at an average speed of 45 miles per hour. How many hours did the trip take him? Remember, *per* means *each* or *one*.

$$\frac{\square \text{ \_\_\_\_\_\_}}{\square \text{ \_\_\_\_\_\_}} = \frac{\square \text{ \_\_\_\_\_\_}}{\square \text{ \_\_\_\_\_\_}}$$

3. Rocky got 2 hits for every 7 times at bat. He batted 56 times. How many hits did Rocky get?

$$\frac{\square \text{ \_\_\_\_\_\_}}{\square \text{ \_\_\_\_\_\_}} = \frac{\square \text{ \_\_\_\_\_\_}}{\square \text{ \_\_\_\_\_\_}}$$

4. 6 pounds of potatoes cost $1.50. How much will 13 pounds of potatoes cost?

$$\frac{\square \text{ \_\_\_\_\_\_}}{\square \text{ \_\_\_\_\_\_}} = \frac{\square \text{ \_\_\_\_\_\_}}{\square \text{ \_\_\_\_\_\_}}$$

Check Your Proportions

Write a proportion. Check each proportion to show that it is true.

If 2 eggs are needed to make 24 cookies, then 5 eggs will be needed to make 60 cookies.

Proportion

$$\frac{2 \text{ eggs}}{24 \text{ cookies}} = \frac{5 \text{ eggs}}{60 \text{ cookies}}$$

Check

$$\frac{2}{24} = \frac{5}{60}$$

$$2 \times 60 = 5 \times 24$$

$$120 = 120$$

Write proportions and check each answer.

1. 3 yards of material cost $5.94, so 2 yards of material will cost $3.96.

   $\frac{\square \text{ \_\_\_\_}}{\square \text{ \_\_\_\_}} = \frac{\square \text{ \_\_\_\_}}{\square \text{ \_\_\_\_}}$ Check

2. Mr. Smith harvested 255 bushels of corn from 3 acres of land. At the same rate, he will harvest 1,700 bushels of corn from 20 acres.

   $\frac{\square \text{ \_\_\_\_}}{\square \text{ \_\_\_\_}} = \frac{\square \text{ \_\_\_\_}}{\square \text{ \_\_\_\_}}$ Check

3. If 3 pounds of fertilizer cover 900 square feet, then 7 pounds of fertilizer will cover 2,100 square feet.

   $\frac{\square \text{ \_\_\_\_}}{\square \text{ \_\_\_\_}} = \frac{\square \text{ \_\_\_\_}}{\square \text{ \_\_\_\_}}$ Check

4. Annette rode her bicycle 120 miles in 3 days. She bicycled 7 days at the same rate and traveled 280 miles.

   $\frac{\square \text{ \_\_\_\_}}{\square \text{ \_\_\_\_}} = \frac{\square \text{ \_\_\_\_}}{\square \text{ \_\_\_\_}}$ Check

Using Proportions

A sports car owner gets 120 miles on 8 gallons of gasoline. How many miles can he travel on 28 gallons?

Proportion

$$\frac{120 \text{ miles}}{8 \text{ gallons}} = \frac{n \text{ miles}}{28 \text{ gallons}}$$

Solution

$$\frac{120}{8} = \frac{n}{28}$$

$$120 \times 28 = n \times 8$$

$$3{,}360 = n \times 8$$

$$n = 3{,}360 \div 8$$

$$n = 420 \text{ miles}$$

Use the proportions and solve each problem.

1. Boys outnumber girls by 3 to 2. If there are 56 girls, how many boys are there?

   Proportion

   $$\frac{3 \text{ boys}}{2 \text{ girls}} = \frac{n \text{ boys}}{56 \text{ girls}}$$

   Solution

2. Roast beef requires a cooking time of 25 minutes per pound. How many minutes would be required to cook a 6-pound roast?

   Proportion

   $$\frac{25 \text{ minutes}}{1 \text{ pound}} = \frac{n \text{ minutes}}{6 \text{ pounds}}$$

   Solution

3. Gilbert's car used 13 gallons of gasoline on a 273-mile trip. How many miles per gallon did his car get?

   Proportion

   $$\frac{13 \text{ gallons}}{273 \text{ miles}} = \frac{1 \text{ gallon}}{n \text{ miles}}$$

   Solution

4. Danna saved $245 in 5 weeks. At the same rate, how much will she save in 7 weeks?

   Proportion

   $$\frac{245 \text{ dollars}}{5 \text{ weeks}} = \frac{n \text{ dollars}}{7 \text{ weeks}}$$

   Solution

Unit Prices

Grocery stores often provide unit prices so that shoppers can compare different sizes of the same item. To find unit prices, you can use proportions.

The unit price is the cost for one item or unit.

A 12-ounce can of orange juice costs $1.68.

Find the unit price.

$$\frac{12 \text{ ounces}}{\$1.68 \text{ cost}} = \frac{1 \text{ ounce}}{n \text{ cost}}$$

$$12 \times n = 1 \times 1.68$$
$$n = 1.68 \div 12$$
$$n = .14$$

The unit price is $.14 per ounce.

$1.44 for 6 pounds of bananas.

Find the unit price.

$$\frac{\$1.44 \text{ cost}}{6 \text{ pounds}} = \frac{n \text{ cost}}{1 \text{ pound}}$$

$$1.44 \times 1 = n \times 6$$
$$1.44 \div 6 = n$$
$$.24 = n$$

The unit price is $.24 per pound.

Find the unit prices.

Dishwashing Detergent

1. 32 ounces for $2.56

$$\frac{\square \text{ ounces}}{\square \text{ cost}} = \frac{\square \text{ ounce}}{\square \text{ cost}}$$

The unit price is ______ per ounce.

Aspirin Tablets

3. $2.64 for 24 tablets

$$\frac{\square \text{ \_\_\_\_\_\_}}{\square \text{ \_\_\_\_\_\_}} = \frac{\square \text{ \_\_\_\_\_\_}}{\square \text{ \_\_\_\_\_\_}}$$

The unit price is ______ per tablet.

Orange Juice

2. 6 ounces for $1.32

$$\frac{\square \text{ \_\_\_\_\_\_}}{\square \text{ \_\_\_\_\_\_}} = \frac{\square \text{ \_\_\_\_\_\_}}{\square \text{ \_\_\_\_\_\_}}$$

The unit price is ______ per ounce.

Toothpaste

4. $1.96 for 7 ounces

$$\frac{\square \text{ \_\_\_\_\_\_}}{\square \text{ \_\_\_\_\_\_}} = \frac{\square \text{ \_\_\_\_\_\_}}{\square \text{ \_\_\_\_\_\_}}$$

The unit price is ______ per ounce.

Find the Unit Costs

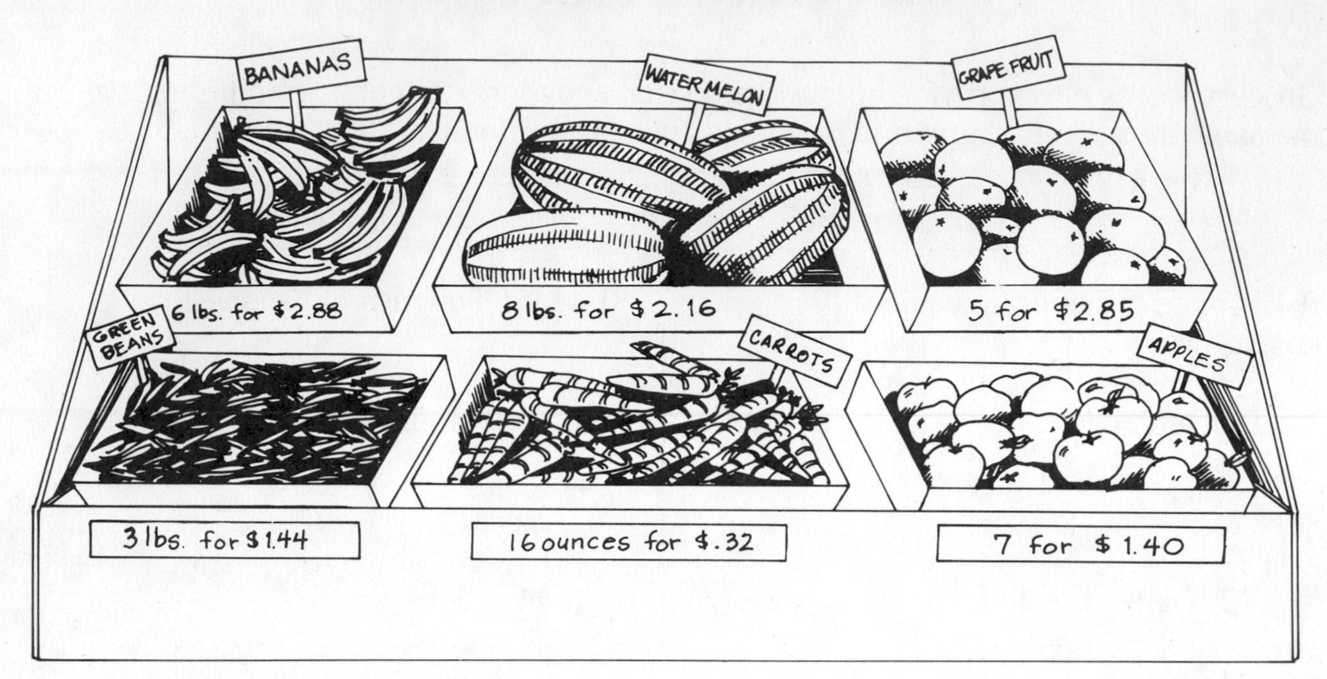

Use proportions to find the unit price of each item.

1. Bananas

$$\frac{6 \text{ pounds}}{2.88 \text{ dollars}} = \frac{1 \text{ pound}}{n \text{ dollars}}$$

Unit price: $ ______ per pound

2. Watermelon

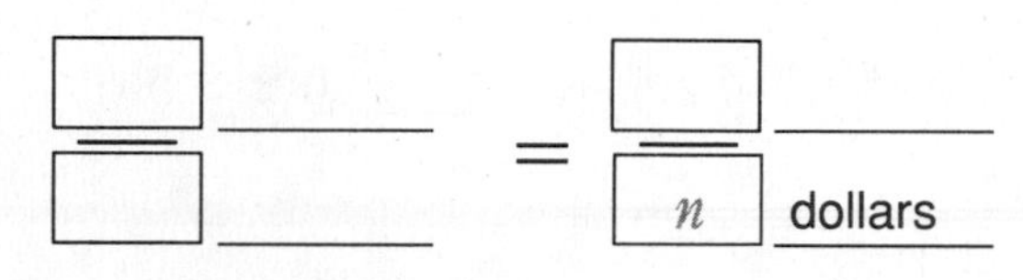

Unit price: $ ______ per pound

3. Grapefruit

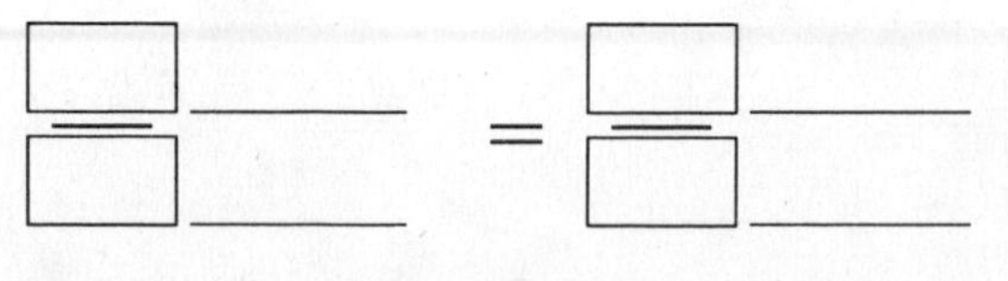

Unit price: $ ______ each

4. Green Beans

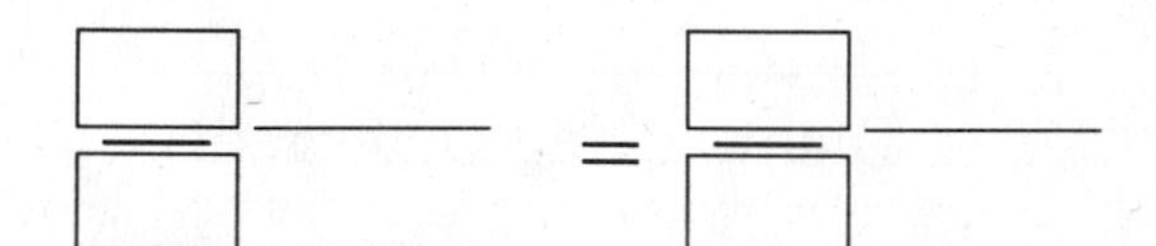

Unit price: $ ______ per pound

5. Carrots

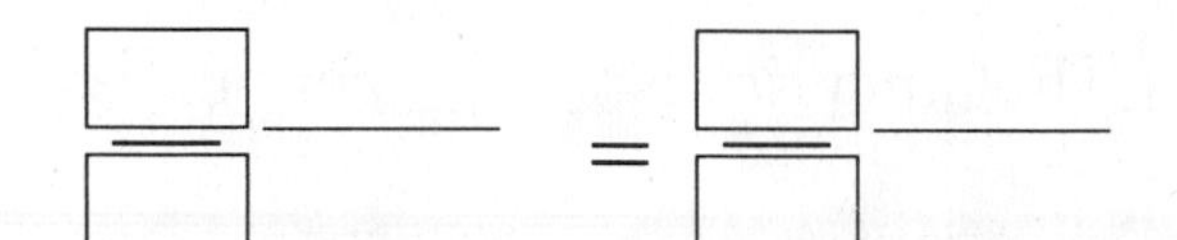

Unit price: $ ______ per ounce

6. Apples

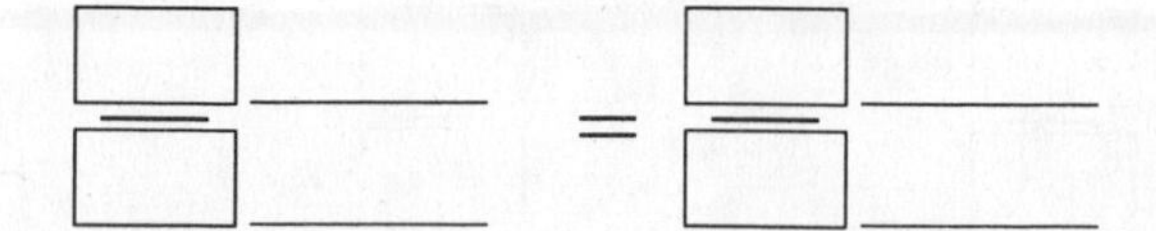

Unit price: $ ______ each

Comparison Shopping

Set up proportions and find the unit prices for two different sizes of a product. Compare the unit prices.

Shampoo

1. 9 ounces for $2.70

$\dfrac{9 \text{ ounces}}{2.70 \text{ dollars}} = \dfrac{1 \text{ ounce}}{n \text{ dollars}}$

The unit price is ______ per ounce.

15 ounces for $3.75

$\dfrac{\square \ \_\_\_\_}{\square \ \_\_\_\_} = \dfrac{1 \text{ ounce}}{n \text{ dollars}}$

The unit price is ______ per ounce.

Which size is cheaper per ounce? ________________

Candy Bars

2. 6 for $2.28

$\dfrac{\square \ \_\_\_\_}{\square \ \_\_\_\_} = \dfrac{\square \ \_\_\_\_}{\square \ \_\_\_\_}$

The unit price is ______ per candy bar.

3 for $1.08

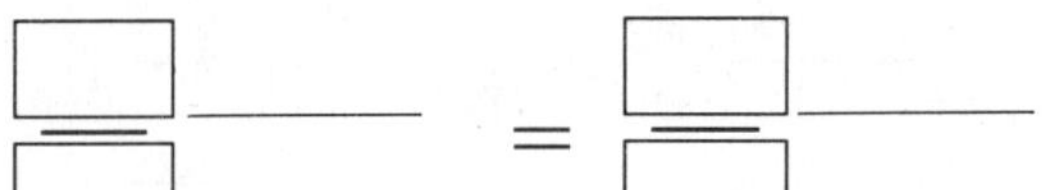

The unit price is ______ per candy bar.

Which package of candy bars is cheaper per bar? ________________

Aluminum Foil

3. 200 feet for $6.00

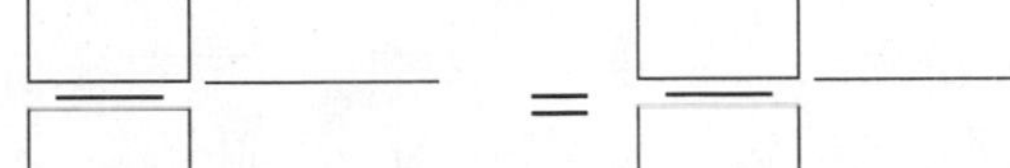

The unit price is ______ per foot.

25 feet for $1.00

$\dfrac{\square \ \_\_\_\_}{\square \ \_\_\_\_} = \dfrac{\square \ \_\_\_\_}{\square \ \_\_\_\_}$

The unit price is ______ per foot.

Which size is cheaper per foot? ________________

Potato Chips

4. 16 ounces for $2.56

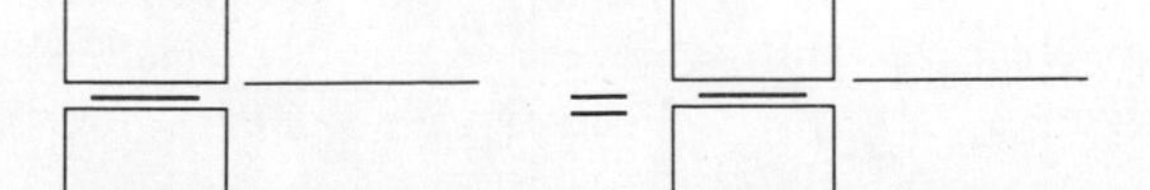

The unit price is ______ per ounce.

6 ounces for $1.32

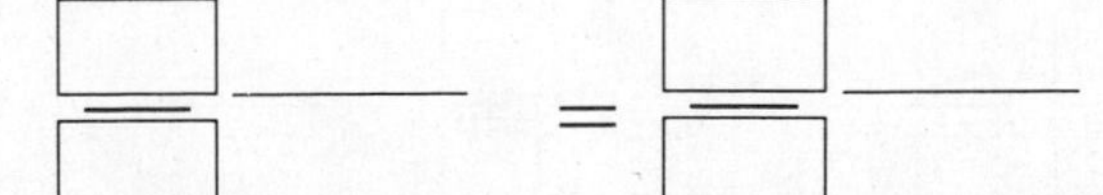

The unit price is ______ per ounce.

Which size is cheaper per ounce? ________________

Changing Recipes

3 eggs are needed to make 15 waffles. How many eggs are needed to make 25 waffles?

WAFFLES

3 eggs

A. $\frac{3}{15}\frac{\text{eggs}}{\text{waffles}} = \frac{n}{25}\frac{\text{eggs}}{\text{waffles}}$

$$3 \times 25 = n \times 15$$
$$75 = n \times 15$$
$$75 \div 15 = n$$
$$\_\_\_\_ = n$$

_____ eggs are needed to make 25 waffles.

1. 2 eggs are needed to make 24 cookies. How many eggs are needed to make 60 cookies?

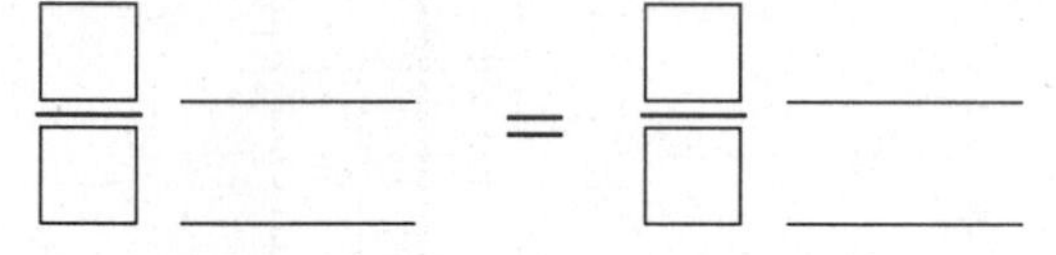

_____ eggs are needed to make 60 cookies.

2. If a recipe uses 3 ounces of cream cheese to make 9 servings, then 4 ounces of cream cheese will make how many servings?

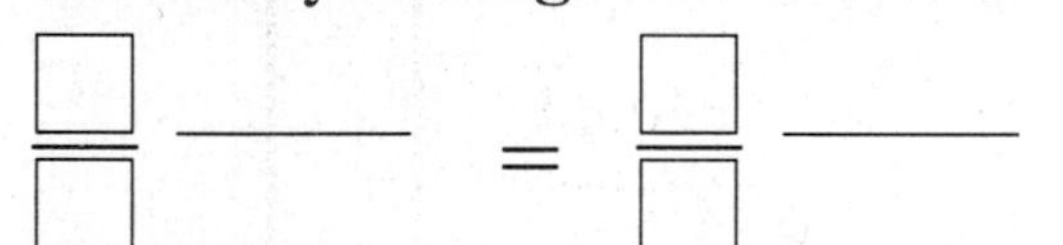

4 ounces of cream cheese will make _____ servings.

3. A fruit punch recipe that serves 12 people calls for 4 oranges. How many oranges are needed to make punch for 18 people?

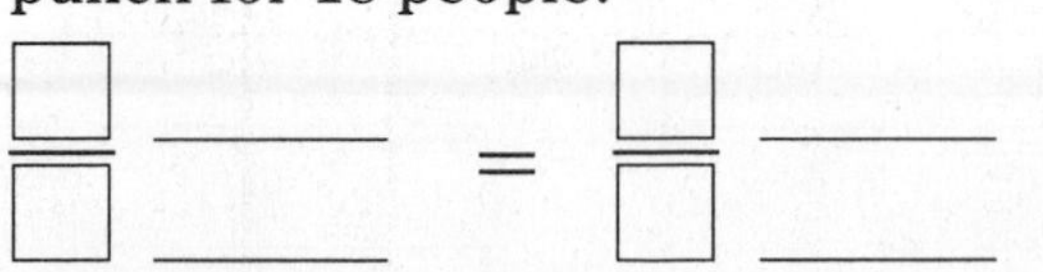

_____ oranges are needed to make punch for 18 people.

4. A soup recipe uses 3 teaspoons butter for 36 servings. How much butter is needed for 12 servings?

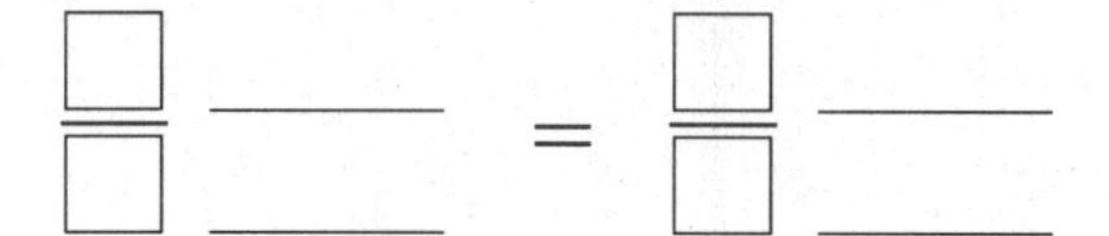

_____ teaspoon of butter is needed for 12 servings.

5. A pancake recipe calls for 6 tablespoons of milk for a serving of 4. How many servings will 9 tablespoons of milk make?

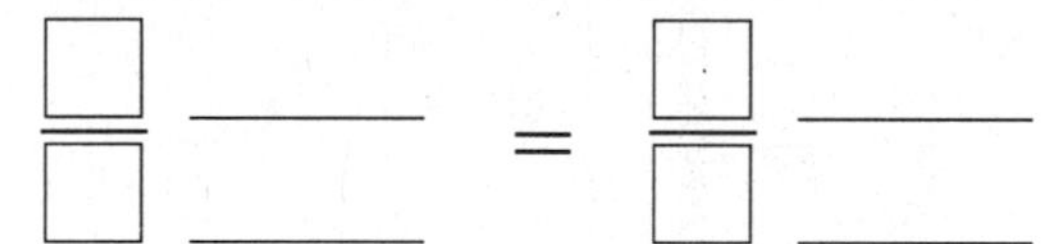

9 tablespoons of milk will make _____ servings.

6. In a biscuit recipe 6 teaspoons of baking powder are needed to make 30 biscuits. How many teaspoons of baking powder are needed to make 40 biscuits?

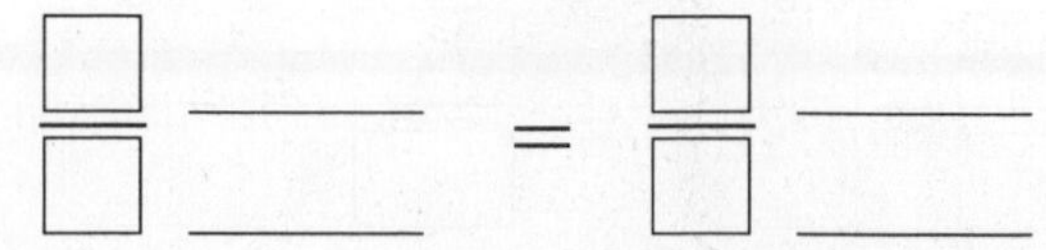

_____ teaspoons of baking powder are needed to make 40 biscuits.

Figuring Costs

Two boxes of cereal cost \$3.96. How much do five boxes of cereal cost?

A. $\frac{2}{3.96}\ \frac{\text{boxes}}{\text{dollars}} = \frac{5}{n}\ \frac{\text{boxes}}{\text{dollars}}$

$2 \times n = 5 \times \$3.96$

$2 \times n = \$19.80$

$n = \$19.80 \div 2$

$n = \$$ ______

Five boxes of cereal cost \$ ______.

To figure the costs, fill in the proportions and solve on another sheet of paper.

1. Material costs \$4.80 per square yard. How much will 8 square yards cost?

   $\frac{\square}{\square}$ dollars / sq. yards $=$ $\frac{\square}{\square}$ dollars / sq. yards

   8 square yards will cost \$ ______.

2. Patricia bought 2 notebooks for \$2.34. How many could she buy for \$7.02?

   $\frac{\square}{\square}$ ______ / ______ $=$ $\frac{\square}{\square}$ ______ / ______

   Patricia could buy ______ notebooks for \$7.02.

3. Craig bought 3 candy bars for \$2.25. How much would 10 candy bars cost?

   $\frac{\square}{\square}$ ______ / ______ $=$ $\frac{\square}{\square}$ ______ / ______

   10 candy bars would cost \$ ______.

4. 6 pounds of potatoes cost \$1.50. How much will 13 pounds of potatoes cost?

   $\frac{\square}{\square}$ ______ / ______ $=$ $\frac{\square}{\square}$ ______ / ______

   13 pounds of potatoes will cost \$ ______.

5. Sherlene bought 9 cans of soup that were marked 3 for \$1.26. How much did she spend?

   $\frac{\square}{\square}$ ______ / ______ $=$ $\frac{\square}{\square}$ ______ / ______

   Sherlene spent \$ ______ for 9 cans of soup.

6. If oranges cost \$1.59 for 3, how much will 7 oranges cost?

   $\frac{\square}{\square}$ ______ / ______ $=$ $\frac{\square}{\square}$ ______ / ______

   7 oranges will cost \$ ______.

Travel Plans

Kevin drives at an average speed of 55 miles per hour. How long will it take him to travel 385 miles?

A. $\frac{55}{1} \frac{\text{miles}}{\text{hour}} = \frac{385}{n} \frac{\text{miles}}{\text{hour}}$

$55 \times n = 385 \times 1$

$55 \times n = 385$

$n = 385 \div 55$

$n = $ ______

It will take Kevin ______ hours.

Use proportions to solve these travel problems.

1. A car averages 21 miles per gallon of gasoline. At this rate, how many gallons will be used on a 315-mile trip?

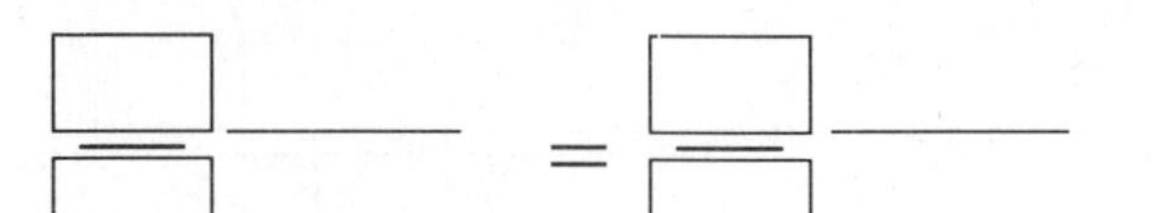

______ gallons of gasoline will be used on a 315-mile trip.

2. A jet flies 1,365 miles in 3 hours. At the same rate of speed, how far can it fly in 5 hours?

The jet can fly ______ miles in 5 hours.

3. If a train can average 55 miles per hour, how many hours would it take the train to travel 220 miles?

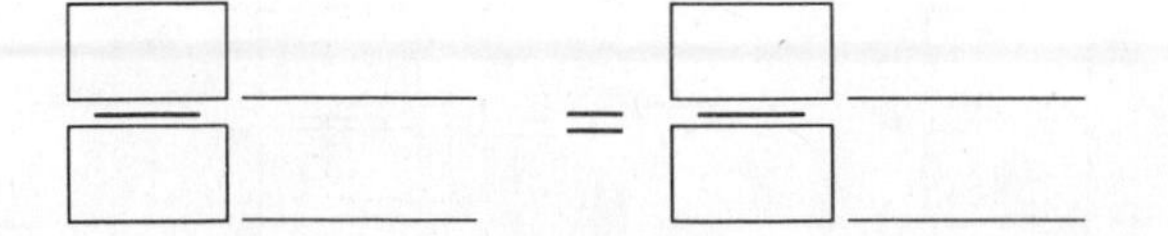

It would take the train ______ hours to travel 220 miles.

4. A train travels at an average speed of 45 miles per hour. At the same rate, how far will it travel in 4.5 hours?

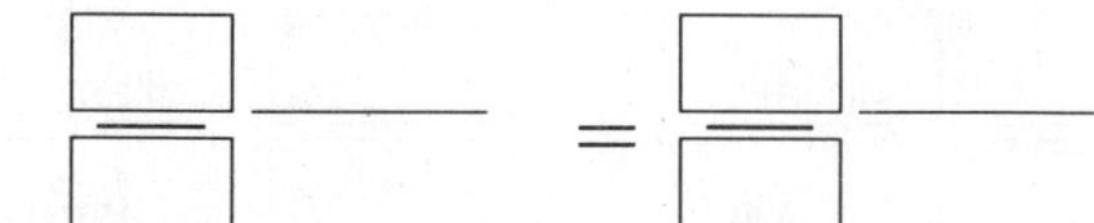

The train will travel ______ miles in 4.5 hours.

5. Carla paid $.35 to travel 25 miles on the toll road. At that rate, how much would it cost to travel 155 miles?

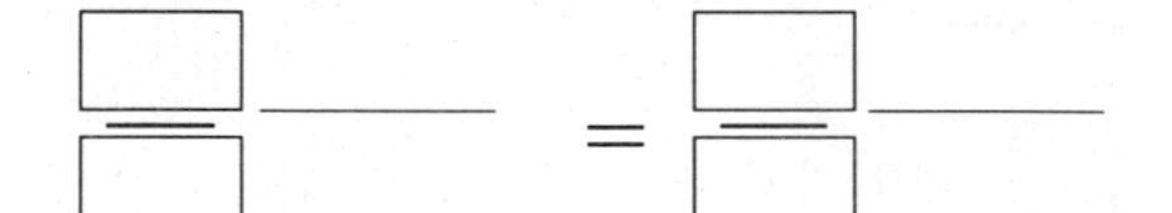

It would cost ______ to travel 155 miles.

6. Benjamin traveled 104 miles in 2 hours. At this rate, how long will it take him to travel 260 miles?

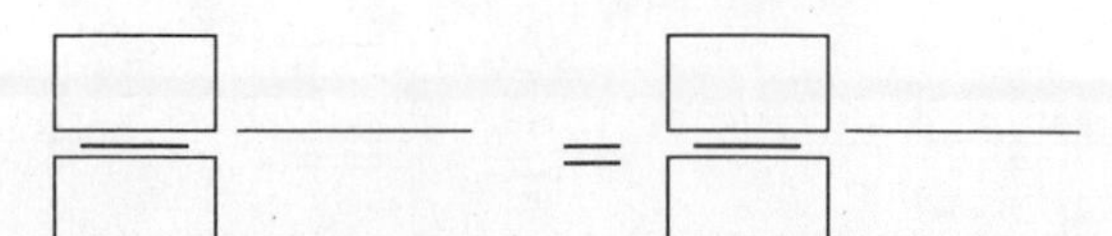

It will take Benjamin ______ hours to travel 260 miles.

Scale Drawings

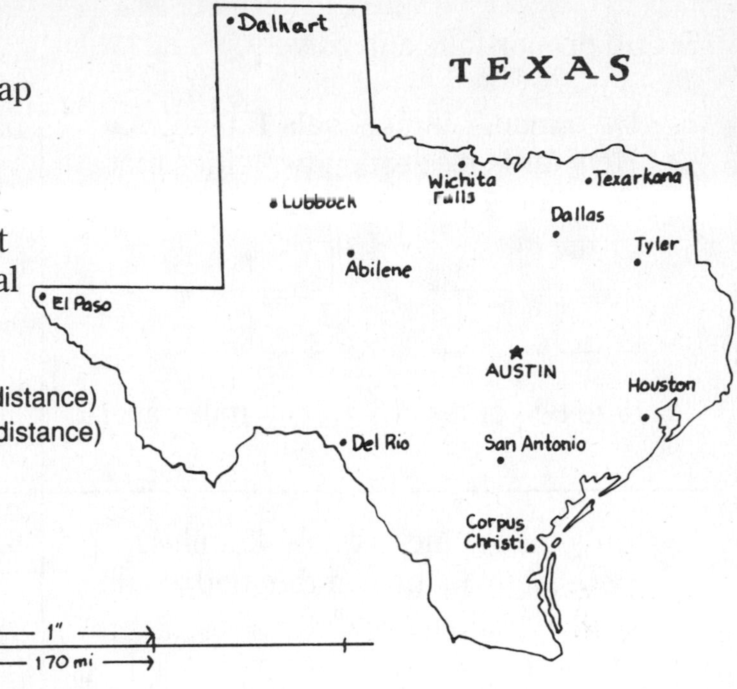

The scale (1 inch = 170 miles) for the Texas map is a ratio that compares the map distance to the actual distance.

1 inch on the map = **170 actual miles.**

The distance on the map between Dalhart and Houston is 3 inches. To find the actual distance in miles, set up a proportion.

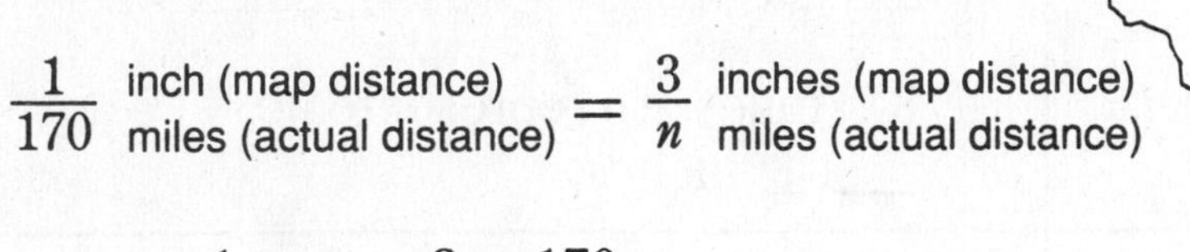

$$\frac{1}{170}\ \frac{\text{inch (map distance)}}{\text{miles (actual distance)}} = \frac{3}{n}\ \frac{\text{inches (map distance)}}{\text{miles (actual distance)}}$$

$$1 \times n = 3 \times 170$$
$$n = 510$$

The actual distance is 510 miles.

Set up proportions and solve. The map distances are given.

1. Find the actual distance between El Paso and San Antonio. The map distance is 2.5 inches.

$$\frac{1}{170}\ \frac{\text{inch (map)}}{\text{miles (actual)}} = \frac{2.5}{n}\ \frac{\text{inches (map)}}{\text{miles (actual)}}$$

The actual distance between El Paso and San Antonio is ______ miles.

2. Find the actual distance between Lubbock and Dallas. The map distance is 1.5 inches.

$$\frac{\square}{\square}\ \frac{\_\_\_\_\_\_}{\_\_\_\_\_\_} = \frac{\square}{\square}\ \frac{\_\_\_\_\_\_}{\_\_\_\_\_\_}$$

The actual distance between Lubbock and Dallas is ______ miles.

3. Find the actual distance between Wichita Falls and Corpus Christi. The map distance is 2 inches.

$$\frac{\square}{\square}\ \frac{\_\_\_\_\_\_}{\_\_\_\_\_\_} = \frac{\square}{\square}\ \frac{\_\_\_\_\_\_}{\_\_\_\_\_\_}$$

The actual distance between Wichita Falls and Corpus Christi is ______ miles.

4. Find the actual distance between Texarkana and Tyler. The map distance is .5 of an inch.

$$\frac{\square}{\square}\ \frac{\_\_\_\_\_\_}{\_\_\_\_\_\_} = \frac{\square}{\square}\ \frac{\_\_\_\_\_\_}{\_\_\_\_\_\_}$$

The actual distance between Texarkana and Tyler is ______ miles.

Map Applications

Set up proportions and solve.

1. On a map, 1 inch equals 150 miles. How far is it between two cities that are 4 inches apart on the map?

   $\frac{\square}{\square}$ ______ = $\frac{\square}{\square}$ ______

   The two cities are ______ miles apart.

2. On a map, 1 inch equals 45 miles. How many inches on the map would represent 315 miles?

   $\frac{\square}{\square}$ ______ = $\frac{\square}{\square}$ ______

   ______ inches on the map represent 315 miles.

3. If a distance of 1 inch on a map equals 75 miles, what actual distance does 3 inches represent?

   $\frac{\square}{\square}$ ______ = $\frac{\square}{\square}$ ______

   3 inches on the map represent ______ miles.

4. On a map, 2 inches equal 150 miles. What actual distance does 5 inches represent?

   $\frac{\square}{\square}$ ______ = $\frac{\square}{\square}$ ______

   5 inches on the map represent ______ miles.

5. On a map of the United States, 2 inches equals 150 miles. How many inches will represent 375 miles?

   $\frac{\square}{\square}$ ______ = $\frac{\square}{\square}$ ______

   375 miles are represented by ______ inches.

6. If 1.5 inches on a map equal 20 miles, how many miles will 6 inches represent?

   $\frac{\square}{\square}$ ______ = $\frac{\square}{\square}$ ______

   6 inches on the map represent ______ miles.

7. On a map, each inch equals 150 miles. How far is it between two cities that are 3.5 inches apart?

   $\frac{\square}{\square}$ ______ = $\frac{\square}{\square}$ ______

   It is ______ miles between two cities that are 3.5 inches apart.

8. A scale drawing shows $\frac{1}{2}$ inch (think .5) equals 50 miles. 7 inches on the scale equals how many miles?

   $\frac{\square}{\square}$ ______ = $\frac{\square}{\square}$ ______

   7 inches on the scale equal ______ miles.

Make a Chart

Read the problem and decide what is being compared.

A recipe for salad dressing calls for mixing 5 tablespoons of oil to 2 tablespoons of vinegar. How many tablespoons of oil should you mix with 5 tablespoons of vinegar?

Compare oil to vinegar.

You can make a chart to organize your information.

oil	5	n
vinegar	2	5

Make a chart for the information. **Do not solve** for n.

1. Alice sleeps nine hours a day. How many hours does she sleep in a year? (Hint: 365 days in a year.)

hours of sleep		
days		

2. For every $10 that Ellen earns, she saves $1. How much would Ellen save if she earned $155?

3. On December 23, it snowed 1.5 inches an hour for 4 hours. How much did it snow in all?

4. At Dan's school, 6 out of every 10 kids know how to swim. If there are 350 kids in his school, how many kids know how to swim?

Turn Charts into Proportions

Read the problem.

For every 2 pounds, roast beef requires a cooking time of 25 minutes. How many minutes are needed to cook a 6-pound roast?

pounds	2	6
minutes	25	n

The chart helps you set up your proportion.

$$\frac{2}{25} = \frac{6}{n}$$

$$2 \times n = 6 \times 25$$

$$2 \times n = 150$$

$$n = 150 \div 2 = 75 \text{ minutes (1 hour and 15 minutes)}$$

Make a chart and write a proportion for each problem. Solve the proportion on another sheet of paper.

1. Larry gets a dollar bill for every 10 dimes that he takes to the bank. How many dollar bills will he get for 130 dimes?

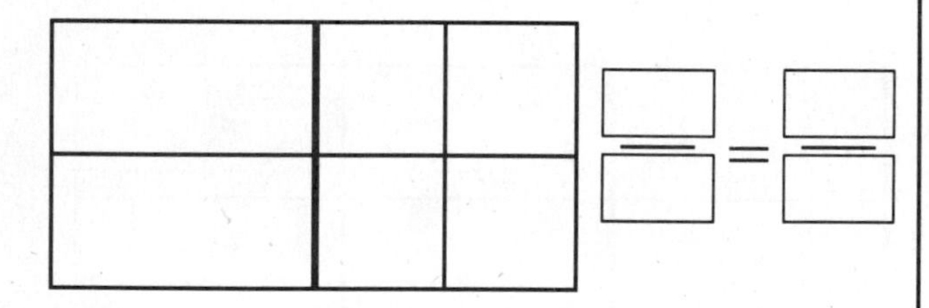

$n =$ ______

2. For every 6 tapes that you buy, you get 2 tapes free. If you bought nine tapes, how many free tapes would you get?

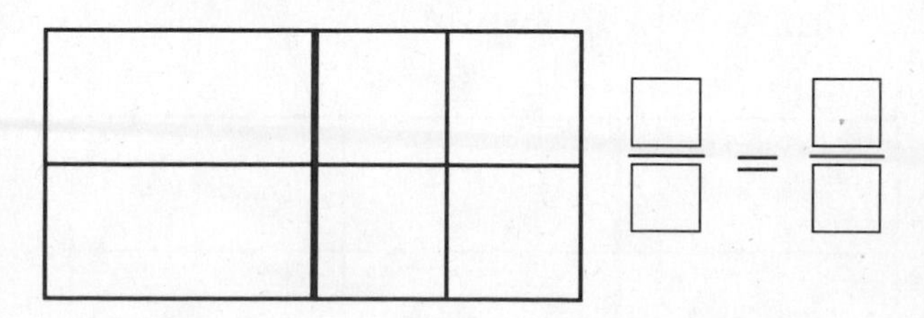

$n =$ ______

3. The snack mix recipe calls for 1.5 boxes of cereal for every bag of pretzels. How many boxes of cereal are needed if 3 bags of pretzels are being used?

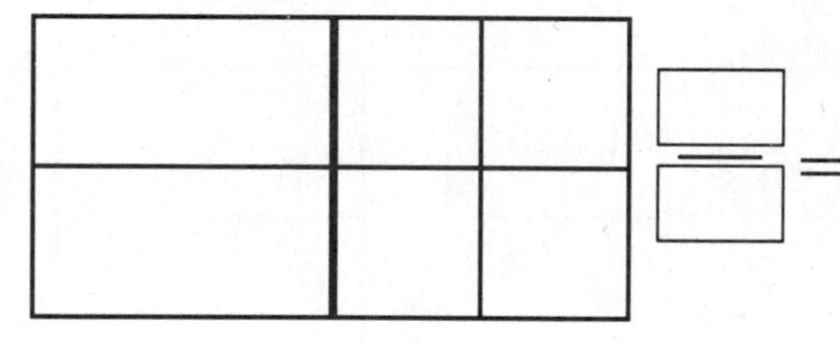

$n =$ ______

4. Janet has a cold. She can take 2 aspirin tablets every 4 hours. How many tablets can she take in 24 hours?

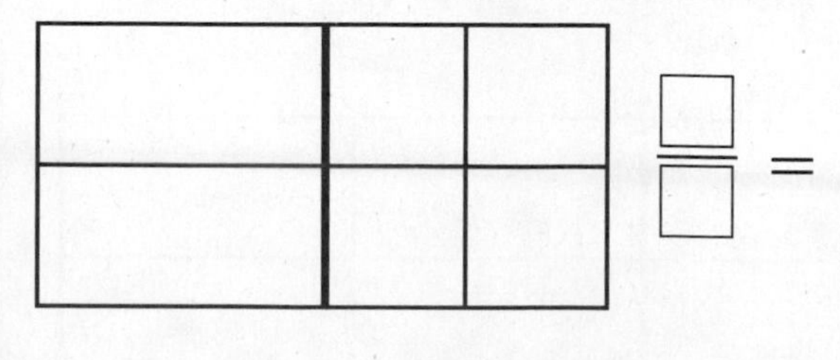

$n =$ ______

Proportion in Measurement

3 feet = 1 yard

How many feet are in 6.5 yards?

Set Up a Chart

feet	3	n
yards	1	6.5

Set Up a Proportion

$$\frac{3 \text{ feet}}{1 \text{ yard}} = \frac{n \text{ feet}}{6.5 \text{ yards}}$$

$$3 \times 6.5 = n \times 1$$

$$19.5 = n$$

There are 19.5 feet in 6.5 yards.

Use a chart to organize the information. Write a proportion and solve on another sheet of paper.

1. 1 pound = 16 ounces. How many pounds are in 128 ounces?

pounds		
ounces		

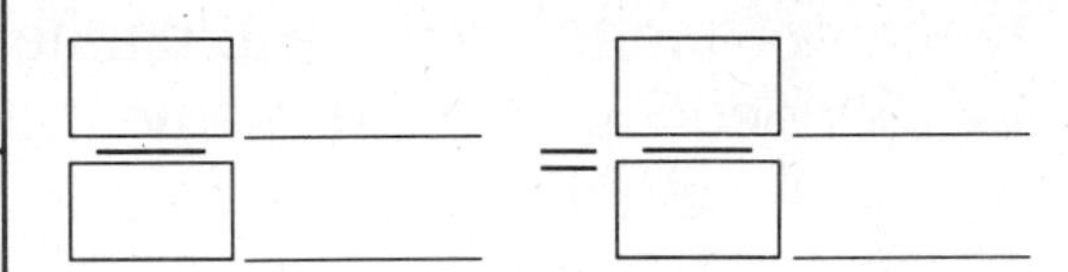

n = ______

2. There are 4 pecks per bushel. How many pecks are in 52 bushels?

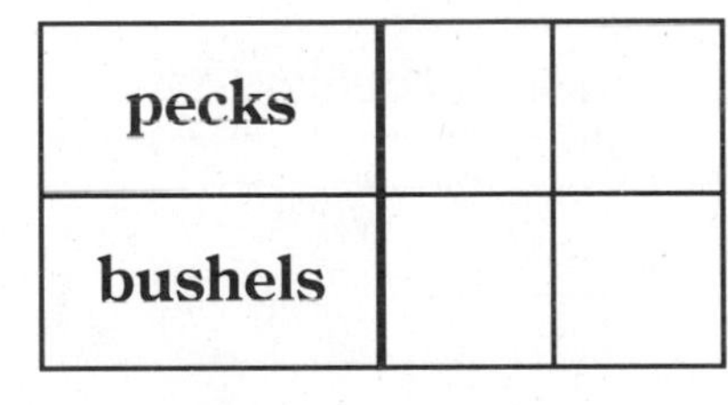

pecks		
bushels		

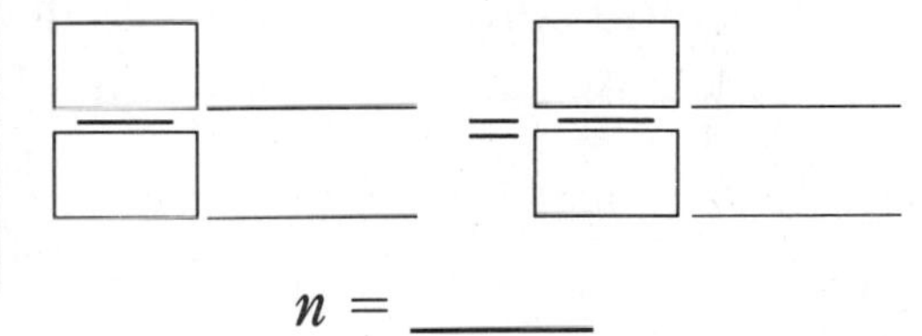

n = ______

3. 1 gram = .04 ounce. How many ounces are in 564 grams?

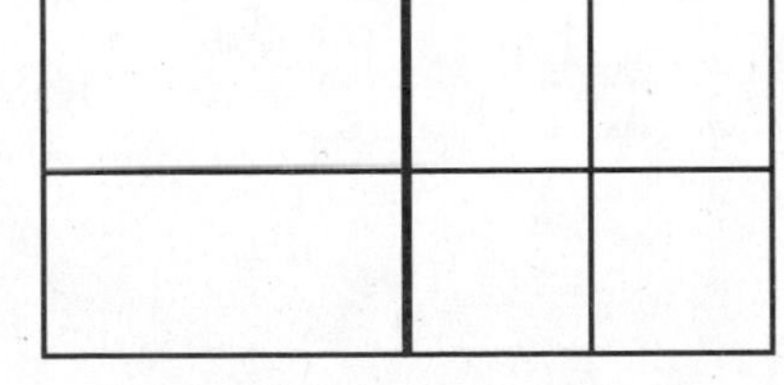

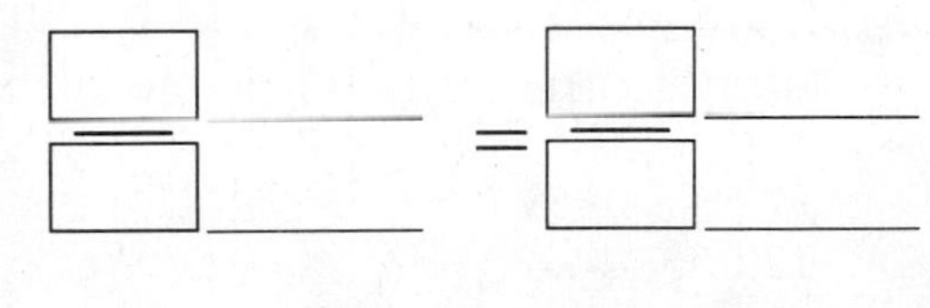

n = ______

4. 100 centimeters = 1 meter. How many centimeters are in 655 meters?

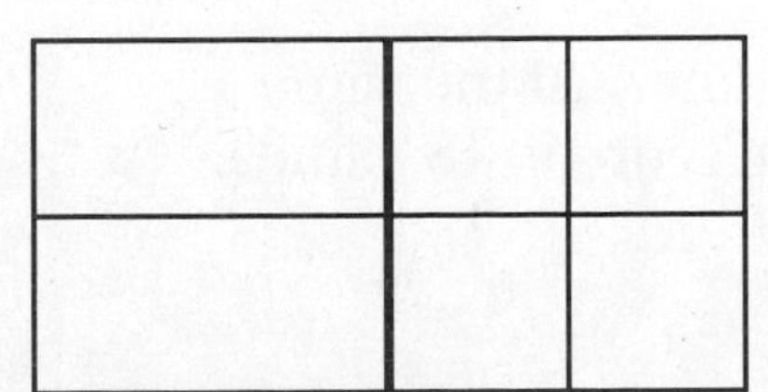

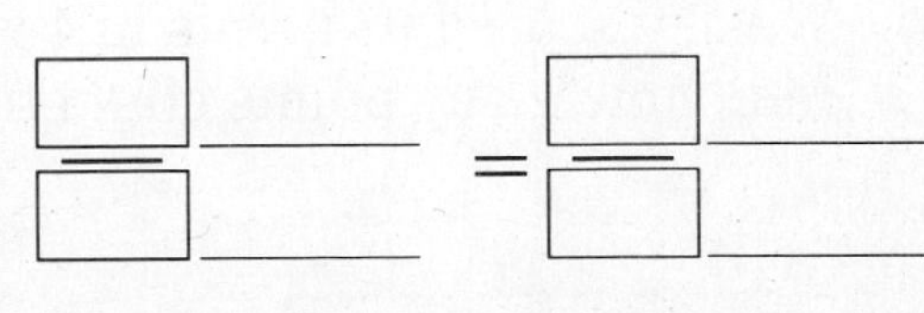

n = ______

Proportions in Sports

Daniel runs 3 miles in 21 minutes. How long does it take him to run 8 miles at the same rate?

Set Up a Chart

miles	3	8
minutes	21	n

Write a Proportion

$$\frac{3}{21} = \frac{8}{n}$$

$$3 \times n = 8 \times 21$$
$$3 \times n = 168$$
$$n = 168 \div 3$$
$$n = 56 \text{ minutes}$$

Use a chart to organize the information. Write a proportion and solve on another sheet of paper.

1. Dora rode her bicycle 120 miles in 3 days. She bicycled 7 days at the same rate. How far did she travel on her bicycle in 7 days?

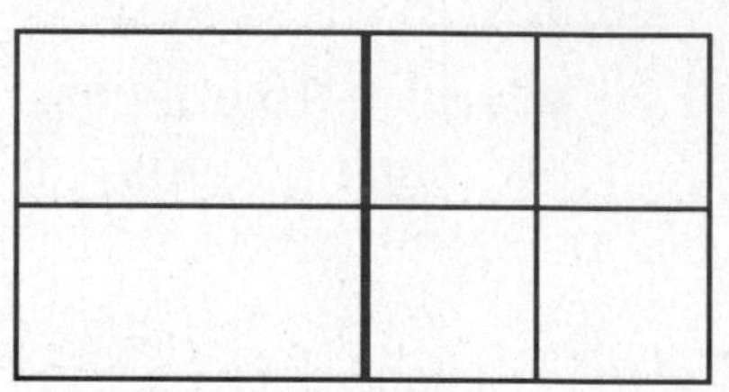

$n =$ ______

2. Edmund can walk 8.4 miles in 2 hours. At the same rate, how far can he walk in 3 hours?

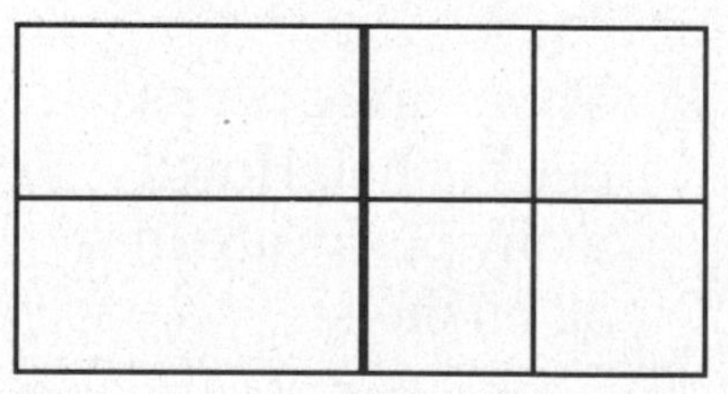

$n =$ ______

3. Mark hit 4 home runs in 6 games. At that rate, how many home runs could he hit in 9 games?

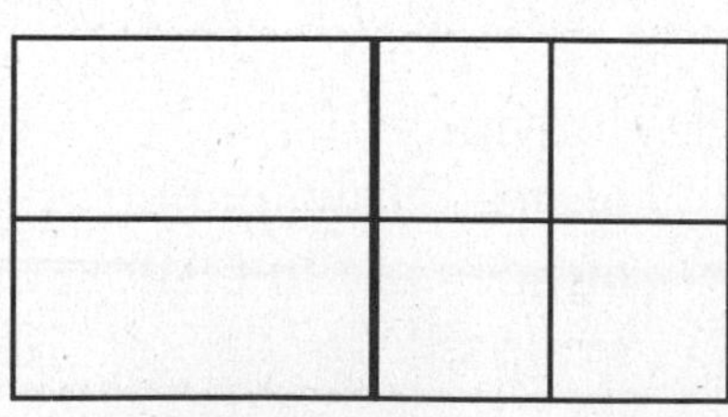

$n =$ ______

4. Wesley scored 92 points in 4 games. At the same rate, how many points could he score in 18 games?

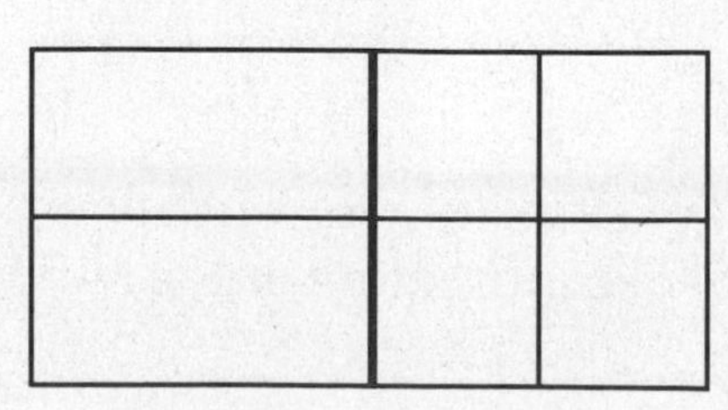

$n =$ ______